IC TIMER COOKBOOK

by
Walter G. Jung

Howard W. Sams & Co., Inc.
4300 WEST 62ND ST. INDIANAPOLIS, INDIANA 46268 USA

$y = \log_a x$

$\frac{\ }{\ } \, y = x$

$_{ga}$

$a^y = x$

$\log_a N = \log_b N \cdot \log_a b$

$= \frac{\log_b N}{\log_b a}$

$2.3026 \log_{10} N = \log_e N$

Preface

Before the 555 timer was introduced in 1972, most monostable and astable RC timing circuits were designed from the ground up, using either discrete semiconductors or ICs as the active elements. The advent of the 555 has changed all of that, making timer designs for most applications a simple hookup of four to five low-cost components. Currently, a 555 (or one of its derivatives) can satisfy about 75% or more of timing requirements with precision, simplicity, and low cost.

Although nothing was revolutionary in its basic concept or theory, the 555 caught on very quickly after its introduction. Perhaps it was just an IC idea whose time had come in terms of getting the right ingredients into a single chip. Nevertheless, its popularity is probably rivaled by no other linear circuit (with the exception of the op amp), and circuit designers are continually producing new and previously undocumented uses for it.

Success, of course, spawns growth in even greater dimensions; the IC timer idea soon became available in duals, then in more sophisticated and wide-range versions, and also in timer/counter combinations. This book discusses all of these types of timers, using representative industry standard devices as examples.

Today the IC timer is king of virtually all RC-timer applications with periods greater than 10 μs, and timing periods can readily be extended into days, weeks, or even months, if desired. The devices are available from many sources, they are versatile, easy to design with, easily programmed or controlled, and interface readily with digital devices. The IC timer represents another milestone of progress in solid-state electronics, and is certainly a tool from which we can all benefit.

Some of the circuits contained in this book have been published previously. In these cases, a specific reference is given accompanying the discussion. The author gratefully acknowledges the work of the authors who have provided the concepts discussed. A general bibliography of timer circuit designs is also contained in Appendix D.

WALTER G. JUNG

Acknowledgments

In the course of preparing this book, the author was aided by various sources of information. For the use of their technical material, the author is grateful to the following companies:

Exar Integrated Systems
Intersil, Inc.
National Semiconductor Corp.
Signetics Corp.

Also, the following industry publications were helpful in allowing use of their copyrighted material:

Electronics
Electronic Design
EDN

The author is also grateful to the following individuals who commented critically on various portions of the manuscript: Bob Zwicker and Alan Grebene of Exar, Bill O'Neil of Intersil, Carl Nelson and Bob Pease of National, and Russ Long and Ted Vaeches of Signetics.

Finally, to my wife Anne goes a special note of thanks for the typing of the manuscript. Thanks also to my research assistants, Jeannie and Mark.

W.G.J.

To Mom

Contents

Part I—Introducing the IC Timer

CHAPTER 1

CHAPTER 2

CHAPTER 3

Part II—IC Timer Applications

CHAPTER 4

CHAPTER 5

CHAPTER 6

IC TIMER SYSTEMS APPLICATIONS 147

Part III—Appendixes

APPENDIX A

MANUFACTURERS' DATA SHEETS 239

APPENDIX B

SECOND-SOURCE GUIDE 267

APPENDIX C

TIMING COMPONENT MANUFACTURERS 271

APPENDIX D

BIBLIOGRAPHY OF IC TIMER DESIGN IDEAS 275

1

INTRODUCING THE IC TIMER

1

RC Timer Basics

The best place to begin a discussion of RC timing circuits is with the basic theory of their operation. By absorbing this fundamental knowledge, we will be better prepared to deal with actual IC devices in practical designs.

1.1 THE MONOSTABLE RC TIMER

The basic operation of a monostable RC timer is illustrated in block diagram form in Fig. 1-1. There are four separate components that make up this timer: a timing resistor (R_t), a timing capacitor (C_t), a switch (S_1), and a threshold/control circuit. The circuit operates as follows:

In its untriggered or quiescent condition, the timer output is low or near zero, and switch S_1 is "on," clamping capacitor C_t to ground. Upon the arrival of a trigger pulse at the input, the control circuit causes switch S_1 to open and causes the output to switch to a high level. The timer is now in its unstable state (i.e., the timing period has begun).

With S_1 now open, capacitor C_t will start to charge through resistor R_t. This causes the voltage across C_t to rise exponentially, forming a timing ramp (see the timing diagram in Fig. 1-1). This voltage continues to rise until it reaches the threshold voltage, V_{th}, which is a voltage that is some fraction of V+. When the timing-ramp voltage reaches V_{th}, the threshold circuit is then reset, the output falls to zero, and the monostable has timed out. The circuit has now returned to its standby (or stable) state.

The output pulse of a monostable timing circuit is said to have a pulse width, T, which is its timing period. This timing period is re-

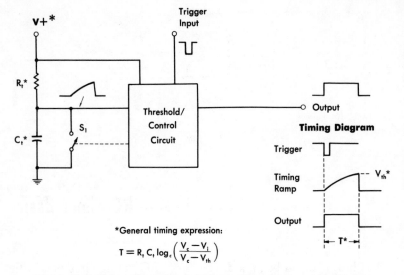

Fig. 1-1. Block diagram illustrating the basic operation of a monostable RC timer.

lated to R_t, C_t, the charging voltage (which in general is V_c; here it is $V+$), and the threshold voltage, V_{th}. The general expression for T is simply

$$T = R_t C_t \log_\epsilon \left(\frac{V_c - V_i}{V_c - V_{th}} \right),$$

where V_c is the charging voltage, and V_i is the initial voltage on C_t. Since C_t starts at an initial voltage of zero, we can simplify the expression for this specific case and also substitute $V+$ for V_c. The new expression is

$$T = R_t C_t \log_\epsilon \left(\frac{V+}{(V+) - V_{th}} \right).$$

In this generalized example we have, for the sake of simplicity, purposely avoided any specific references to actual resistor or capacitor values. And, although an example of a positive source voltage $(V+)$ was shown, in theory we could just as well have a negative supply, a negative ramp, a negative V_{th}, etc.

Fig. 1-2 illustrates how an actual circuit could be connected to implement this monostable timer. Here R_t and C_t are as they were in Fig. 1-1, but S_1 is replaced by a transistor switch, Q_1. The function of Q_1 is to short C_t in the standby state, and to open during the timing period. The control function is implemented by a flip-flop, whose output is to be the monostable timing pulse of width T. This flip-flop also controls transistor Q_1; Q_1 is "on" when the output is low, and is

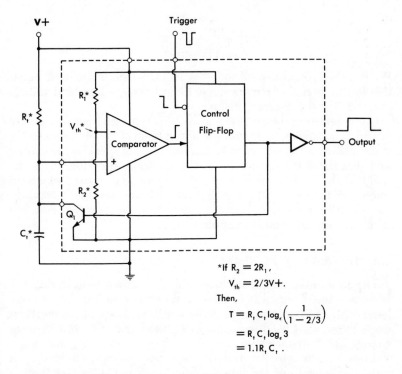

Fig. 1-2. Circuit implementation of the basic monostable RC timer.

"off" when the output is high. The threshold function is performed by the comparator and the voltage divider, R_1–R_2. The voltage divider is connected across V+ (the supply line), and its output voltage is V_{th}, the threshold voltage. The output of the comparator will change states when the timing ramp is equal to V_{th}.

The sequence of operation of this circuit is precisely as described before, but a deeper appreciation of it may be gained from an example of operation. For instance, if resistors R_2 and R_1 are chosen to have a ratio of 2:1 ($R_2 = 2R_1$), the voltage divider output of V_{th} will be 2/3 of V+. With this fact in mind, we can write a specific equation for period T. All the voltages can be expressed in terms of V+, since $V_{th} = 2/3V+$. We then write

$$T = R_t C_t \log_\epsilon \left(\frac{V+}{(V+) - 2/3V+} \right),$$

which simplifies to

$$T = R_t C_t \log_\epsilon \left(\frac{1}{1 - 2/3} \right),$$

or

$$T = R_tC_t\log_\epsilon 3$$
$$= 1.0986R_tC_t.$$

We can then round this off to $T = 1.1R_tC_t$, which is the basic equation for the period of a monostable timer having a threshold voltage that is 2/3 of the charging voltage.

One interesting thing to note about this type of circuit is that the period is not dependent upon the absolute level of V+, because V_{th} is derived as a fraction of V+. Mathematically speaking, it can be said that the V+ term divides out. Electrically speaking, it can be said that the capacitor is both charged from and compared to a fixed fraction of the same voltage. Either way it is an important advantage (as we will soon see) to have a basic timing equation that does not critically depend upon the supply voltage.

1.2 THE ASTABLE RC TIMER

The astable RC timer is shown in block diagram form in Fig. 1-3. Note that this diagram is in many ways similar to Fig. 1-1; it does, however, have two timing resistors and two threshold connections since there are two threshold levels associated with this type of circuit. The two thresholds are the voltage levels V_{th+} and V_{th-}. Voltage V_{th+} is some fraction of the supply voltage, while V_{th-} is a smaller fraction of the same supply voltage. This is just another way of saying that V_{th+} is more positive than V_{th-}, a necessary condition for circuit operation. Operation of the circuit is as follows:

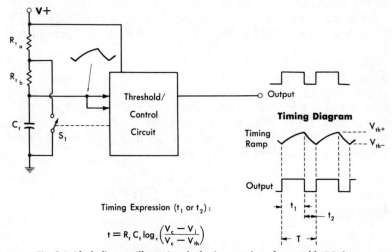

Timing Expression (t_1 or t_2):

$$t = R_tC_t\log_\epsilon\left(\frac{V_c - V_i}{V_c - V_{th}}\right)$$

Fig. 1-3. Block diagram illustrating the basic operation of an astable RC timer.

At the beginning, assume switch S_1 to be open. In this state, the output will be high and capacitor C_t will charge toward V+ through resistors R_{t_a} and R_{t_b}. This portion of the cycle is somewhat similar to that described previously in the case of the monostable timer. When the timing ramp across C_t reaches the voltage level of V_{th+}, the circuit changes states. The output now goes low, which causes switch S_1 to close. With S_1 closed, the R_{t_a}–R_{t_b} node is grounded, which places R_{t_b} in parallel with C_t, and effectively removes R_{t_a} from the circuit. C_t now begins to discharge through R_{t_b}, and the timing ramp decays exponentially toward ground. When the voltage across C_t reaches the lower threshold of V_{th-}, the circuit once again will revert to its high output state, with S_1 opening and C_t charging toward V+.

The circuit will continue to oscillate between the two threshold voltage points of V_{th+} and V_{th-}, with the output changing state with each threshold crossing. Referring to the timing diagram in Fig. 1-3, the positive-going timing period is termed t_1, and the output is high during this period. The negative-going timing period is t_2, and the output is low during this period. The total timing period of a single cycle is called T, which is simply the sum of the individual timing periods, t_1 and t_2. For each of the two timing periods, the general expression will again be

$$t = R_t C_t \log_\epsilon \left(\frac{V_c - V_i}{V_c - V_{th}} \right).$$

A circuit implementation of this astable timer is illustrated in Fig. 1-4. Here we note that there are two comparators, the upper comp and the lower comp. These comparators establish the two threshold voltages, V_{th+} and V_{th-}, as the fractions of V+ determined by the divider resistor string, R_1–R_2–R_3. The upper comp is referenced to the higher voltage, V_{th+}. The lower comp is referenced to the lower voltage, V_{th-}. Transistor Q_1 performs a function similar to that of switch S_1 in Fig. 1-3, while the control flip-flop drives Q_1 and the output buffer as directed by the comparator inputs.

In operation, the circuit performs just as described for Fig. 1-3, with an output and timing ramp in accordance with the timing diagram. The voltage across C_t is made to oscillate between the two comparator thresholds, V_{th+} and V_{th-}. $R_{t_a} + R_{t_b}$ and C_t control timing period t_1, while R_{t_b} and C_t control timing period t_2. The general timing expression for either of these periods is as noted in Fig. 1-3. In this case, with comparator thresholds $V_{th+} = 2/3$V+ and $V_{th-} = 1/3$V, the equations for t_1 and t_2 are as follows:

For t_1, the voltage across C_t starts at a voltage of V_{th-} (which is V_i), charges toward V+ (which is V_c), and reaches its upper limit at V_{th+}. Then,

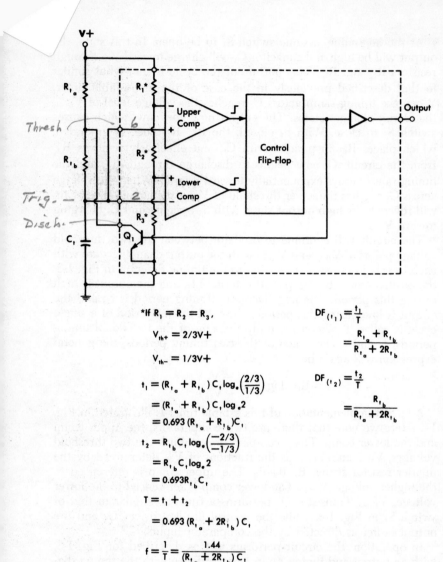

*If $R_1 = R_2 = R_3$,

$$V_{th+} = 2/3V+$$

$$V_{th-} = 1/3V+$$

$$t_1 = (R_{t_a} + R_{t_b}) C_t \log_e \left(\frac{2/3}{1/3}\right)$$

$$= (R_{t_a} + R_{t_b}) C_t \log_e 2$$

$$= 0.693 (R_{t_a} + R_{t_b}) C_t$$

$$t_2 = R_{t_b} C_t \log_e \left(\frac{-2/3}{-1/3}\right)$$

$$= R_{t_b} C_t \log_e 2$$

$$= 0.693 R_{t_b} C_t$$

$$T = t_1 + t_2$$

$$= 0.693 (R_{t_a} + 2R_{t_b}) C_t$$

$$f = \frac{1}{T} = \frac{1.44}{(R_{t_a} + 2R_{t_b}) C_t}$$

$$DF_{(t_1)} = \frac{t_1}{T}$$

$$= \frac{R_{t_a} + R_{t_b}}{R_{t_a} + 2R_{t_b}}$$

$$DF_{(t_2)} = \frac{t_2}{T}$$

$$= \frac{R_{t_b}}{R_{t_a} + 2R_{t_b}}$$

Fig. 1-4. Circuit implementation of the basic astable RC timer.

$$t_1 = (R_{t_a} + R_{t_b}) C_t \log_\epsilon \left(\frac{(V+) - V_{th-}}{(V+) - V_{th+}}\right).$$

With V_{th+} and V_{th-} as fractions of V+ as noted, this may be written as

$$t_1 = (R_{t_a} + R_{t_b}) C_t \log_\epsilon \left(\frac{2/3V+}{1/3V+}\right),$$

which reduces to

$$t_1 = (R_{t_a} + R_{t_b}) C_t \log_e 2$$
$$= 0.693 (R_{t_a} + R_{t_b}) C_t.$$

For t_2, the equations are similar:

$$t_2 = R_{t_b} C_t \log_e \left(\frac{0 - V_{th+}}{0 - V_{th-}}\right),$$

which simplifies to

$$t_2 = R_{t_b} C_t \log_e \left(\frac{-2/3V+}{-1/3V+}\right)$$
$$= R_{t_b} C_t \log_e 2$$
$$= 0.693 R_{t_b} C_t.$$

The total period, T, is simply the sum of periods t_1 and t_2, or

$$T = t_1 + t_2$$
$$= 0.693 (R_{t_a} + R_{t_b}) C_t + 0.693 R_{t_b} C_t$$
$$= 0.693 (R_{t_a} + 2R_{t_b}) C_t.$$

Since time and frequency are related reciprocally, we can now also write an equation for the operating frequency, f:

$$f = \frac{1}{T}$$
$$= \frac{1}{0.693 (R_{t_a} + 2R_{t_b}) C_t}$$
$$= \frac{1.44}{(R_{t_a} + 2R_{t_b}) C_t}.$$

The ratio of the individual periods (t_1 or t_2) to the total period is called the duty factor, DF. With respect to period t_1, the duty factor is

$$DF_{(t_1)} = \frac{t_1}{T}$$
$$= \frac{R_{t_a} + R_{t_b}}{R_{t_a} + 2R_{t_b}},$$

and for period t_2, the duty factor is

$$DF_{(t_2)} = \frac{t_2}{T}$$
$$= \frac{R_{t_b}}{R_{t_a} + 2R_{t_b}}.$$

In summary, there are a number of important features that characterize the astable RC timer. The timing period is governed by C_t and resistors R_{t_a} and R_{t_b}, as is the operating frequency. The duty cycle is controlled by the ratio of timing resistors. And, for this type of timer as well as for the monostable, the timing period is independent of the supply voltage.

What has been described in this chapter is a concept that allows (in theory at least) a very simple and predictable form of monostable and/or astable timer design based only on R and C values to define operation. With this groundwork, we are now ready to examine IC timer types and their modes of operation.

2

IC Timer Types

The first chapter, while not an all-encompassing treatise on RC timing circuits, does provide sufficient background for the understanding of those ICs that use the principles outlined. Specifically, these include the 555 general-purpose timer and also others based either wholly or in part on its operating theory. In this chapter, we begin a more detailed examination of the internal workings of the 555. This examination, in conjunction with the first chapter, will lay a foundation for the entire book.

2.1 THE 555 SINGLE-UNIT GENERAL-PURPOSE TIMER

The NE555 timer, manufactured by Signetics, was not only the first IC timer introduced (in 1972), but also the trendsetter for most of the devices that have been designed since. The 555 is a general-purpose unit, capable of both the monostable and astable operating modes over wide ranges. Actually, the operating principles of the 555 are the same as those described in Chapter 1, with but few differences in practice. Most manufacturers supply the 555 in both the 8-lead TO-99 metal can and the 8-pin dual in-line (MINIDIP) packages, while some manufacturers also offer the unit in a 14-pin dual in-line (DIP) package.

2.1.1 Functional Diagram and Schematic

The 555 timer can be functionally diagrammed as shown in Fig. 2-1. The reader will note a similarity to the circuit diagram of Fig. 1-4. In actuality, the basic theory of the 555 has already been explained. The implementation of the various circuit functions within the device will now be described.

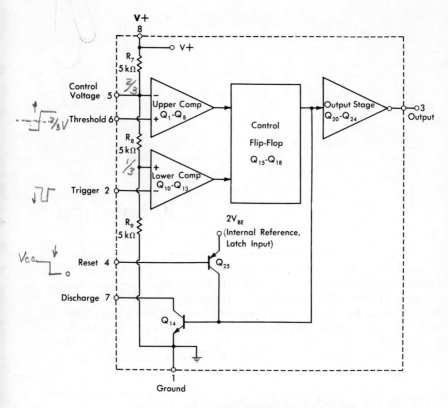

Fig. 2-1. Functional block diagram of the 555 timer.

In Fig. 2-1, note the resistive divider string across the V+ line comprising equal-value resistors R_7, R_8, and R_9. This voltage divider provides the reference voltages for the upper and lower comparators of 2/3V+ and 1/3V+, respectively. In the schematic* of Fig. 2-2, this divider may also be noted, biasing Q_4 and Q_{13}. Transistors Q_1–Q_8 make up the upper comparator, while Q_{10}–Q_{13} form the lower comparator. In both comparators, Darlington differential input stages are used for low (100 nA) input currents, which in turn allows a wide range of (external) timing resistor values to be used. The upper comparator point of 2/3V+ is brought outside the IC package (via pin 5) to allow external control of the timing period (when desired).

---

* Some variation in internal circuitry is evident between different manufacturers of 555 devices; therefore, reference designations and exact circuit details may vary from that shown here when compared with a given data sheet.

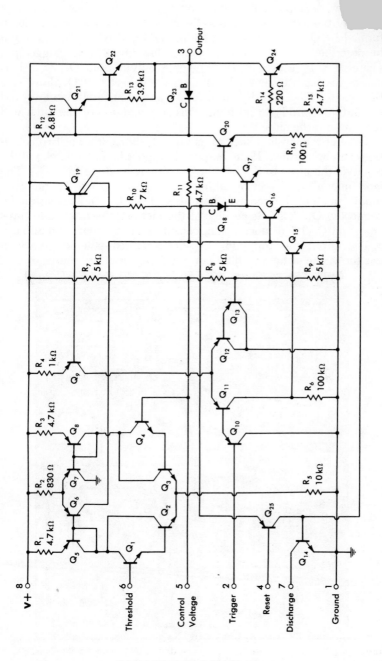

Fig. 2-2. Schematic of the 555 timer.

two comparator outputs are taken from transistors Q_6 and Q_{11}; these in turn are fed to the control flip-flop, which is a latch formed by Q_{16}–Q_{17} (Q_{18} is an additional input whose function will be described shortly). In operation, a low input to the trigger pin (base of Q_{10}) causes a positive-going output from Q_{10}–Q_{11}. This causes the latch to be set by pulling the collector of Q_{15} low. This then causes Q_{17} (and the input to Q_{20} and Q_{24}) to also go low, and the output (pin 3) to go high. This set condition of the latch will remain until the circuit is reset.

To reset the flip-flop back to its original state, either of two conditions can be satisfied. If the output of Q_6 goes high, the latch will be reset via the input to the base of Q_{16}, removing drive from Q_{17}. The latch may also be reset by taking the base of Q_{25} low, via the reset input (pin 4). This removes base drive from Q_{17} by biasing diode-connected Q_{18} "off." Regardless of the method, the reset state turns Q_{20} (and Q_{24}) "on" once again, causing an output-low condition.

The output stage formed by Q_{20}–Q_{24} is a totem-pole design, which has the virtue of being a high current drive for either source or sink

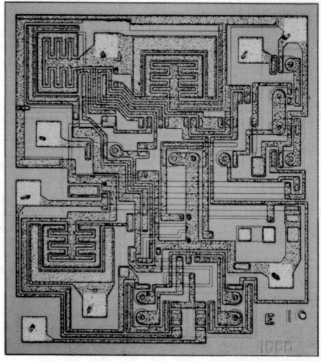

Fig. 2-3. Photomicrograph of the 555 silicon chip.

loads. This design is a versatile one, as it can readily drive TTL inputs with a chip supply of 5 volts, yet it can also sink or source 200 mA when operated from 15 volts. A photomicrograph of the 555 silicon chip layout is shown in Fig. 2-3.

2.1.2 Definition of Pin Functions

As an aid to understanding the 555 more completely, this section provides a short description of the functional characteristics of each pin. This not only will serve as an aid to understanding the use of the 555 as a timer, but will also greatly facilitate its use in some of the more imaginative nontimer uses of which it is capable. To amplify this discussion, the reader is referred to the 555 data sheet reproduced in Appendix A.

V+ (Pin 8)

The V+ pin (referred to as V_{CC} by some manufacturers) is the positive supply voltage terminal of the device. Supply-voltage operating range for the 555 is +4.5 volts (minimum) to +16 volts (maximum), and it is specified for operation between +5 volts and +15 volts. The device will operate essentially the same over this range of voltages without change in timing period. Actually, the most significant operational difference is the output drive capability, which increases for both current and voltage range as the supply voltage is increased. Sensitivity of time interval to supply voltage change is low, typically 0.1% per volt.

Ground (Pin 1)

The ground (or common) pin is the most-negative supply potential of the device, which is normally connected to circuit common when operated from positive supply voltages.

Output (Pin 3)

The output of the 555 comes from a high-current totem-pole stage made up of transistors Q_{20}–Q_{24}. Transistors Q_{21} and Q_{22} provide drive for source-type loads, and their Darlington connection provides a high-state output voltage about 1.7 volts less than the V+ supply level used. Transistor Q_{24} provides current-sinking capability for low-state loads referred to V+ (such as typical TTL inputs). Transistor Q_{24} has a low saturation voltage, which allows it to interface directly, with good noise margin, when driving current-sinking logic. Exact output saturation levels vary markedly with supply voltage, however, for both high and low states. At a V+ of 5 volts, for instance, the low state $V_{CE(sat)}$ is typically 0.25 volt at 5 mA. Operating at 15 volts, however, it can sink 100 mA if an output-low voltage level of 2 volts is allowable (power dissipation should be considered in such a case, of course). High-state level is typically 3.3 volts at V+ = 5 volts; 13.3 volts at V+ = 15 volts. Both the rise and fall times of the output waveform are quite fast, typical switching times being 100 ns.

The state of the output pin will always reflect the inverse of the logic state of the latch, and this fact may be seen by examining Fig. 2-1 (or Fig. 2-2). Since the latch itself is not directly accessible, this relationship may be best explained in terms of latch-input trigger conditions. To trigger the output to a high condition, the trigger input is momentarily taken from a higher to a lower level. [The exact voltage levels are discussed under "Trigger (Pin 2)."] This causes the latch to be set and the output to go high. Actuation of the lower comparator is the only manner in which the output can be placed in the high state. The output can be returned to a low state by causing the threshold to go from a lower to a higher level [exact levels are discussed under "Threshold (Pin 6)"], which resets the latch. The output can also be made to go low by taking the reset to a low state near ground [exact levels are discussed under "Reset (Pin 4)"].

Control Voltage (Pin 5)

This pin allows direct access to the 2/3V+ voltage-divider point, the reference level for the upper comparator. It also allows indirect access to the lower comparator, as there is a 2:1 divider (R_8–R_9) from this point to the lower-comparator reference input, Q_{13}. Use of this terminal is the option of the user, but it does allow extreme flexibility by permitting modification of the timing period, resetting of the comparator, etc.

When the 555 timer is used in a voltage-controlled mode, its voltage-controlled operation ranges from about 1 volt less than V+ down to within 2 volts of ground (although this is not guaranteed). Voltages can be safely applied outside these limits, but they should be confined within the limits of V+ and ground for reliability.

In the event the control-voltage pin is not used, it is recommended that it be bypassed with a capacitor of about 0.01 μF for immunity to noise, since it is a comparator input.

Trigger (Pin 2)

This pin is the input to the lower comparator and is used to set the latch, which in turn causes the output to go high. This is the beginning of the timing sequence in monostable operation. Triggering is accomplished by taking the pin from above to below a voltage level of 1/3V+ (or, in general, one-half the voltage appearing at pin 5). The action of the trigger input is level-sensitive, allowing slow rate-of-change waveforms, as well as pulses, to be used as trigger sources.

One precaution that should be observed with the trigger input signal is that it must not remain lower than 1/3V+ for a period of time *longer* than the timing cycle. If this is allowed to happen, the timer will retrigger itself upon termination of the first output pulse. Thus, when the timer is driven in the monostable mode with input pulses longer than the desired output pulse width, the input trigger should effectively be shortened by differentiation.

The minimum-allowable pulse width for triggering is somewhat dependent upon pulse level, but in general if it is greater than 1 μs, triggering will be reliable (see data sheet in Appendix A).

A second precaution with respect to the trigger input concerns storage time in the lower comparator. This portion of the circuit can exhibit normal turn-off delays of several microseconds after triggering; that is, the latch can still have a trigger input for this period of time *after* the trigger pulse. In practice, this means the minimum monostable output pulse width should be on the order of 10 μs to prevent possible double triggering due to this effect.

The voltage range that can safely be applied to the trigger pin is between V+ and ground. A dc current, termed the *trigger* current, must also flow from this terminal into the external circuit. This current is typically 500 nA and will define the upper limit of resistance allowable from pin 2 to ground. For an astable configuration operating at V+ = 5 volts, this resistance is 3 MΩ; it can be greater for higher V+ levels.

Threshold (Pin 6)

This pin is one input to the upper comparator (the other being pin 5) and is used to reset the latch, which causes the output to go low. Resetting via this terminal is accomplished by taking the terminal from below to above a voltage level of 2/3V+ (the normal voltage on pin 5). The action of the threshold pin is level sensitive, allowing slow rate-of-change waveforms.

The voltage range that can safely be applied to the threshold pin is between V+ and ground. A dc current, termed the *threshold* current, must also flow into this terminal from the external circuit. This current is typically 100 nA, and will define the upper limit of total resistance allowable from pin 6 to V+. For either timing configuration operating at V+ = 5 volts, this resistance is 16 MΩ.

Reset (Pin 4)

This pin is also used to reset the latch and return the output to a low state. The reset voltage threshold level is 0.7 volt, and a sink current of 0.1 mA from this pin is required to reset the device. These levels are relatively independent of operating V+ level; thus the reset input is TTL compatible for any supply voltage.

The reset input is an overriding function; that is, it will force the output to a low state regardless of the state of either of the other inputs. It may thus be used to terminate an output pulse prematurely, to gate oscillations from "on" to "off," etc. Delay time from reset to output is typically on the order of 0.5 μs, and the minimum reset pulse width is 0.5 μs. Neither of these figures are guaranteed, however, and may vary from one manufacturer to another. When not used, it is recommended that the reset input be tied to V+ to avoid any possibility of false resetting.

Discharge (Pin 7)

This pin is the open collector of an npn transistor (Q_{14}, Fig. 2-2), the emitter of which goes to ground. The conduction state of this transistor is identical in timing to that of the output stage. It is "on" (low resistance to ground) when the output is low and "off" (high resistance to ground) when the output is high.

In both the monostable and astable timer modes, this transistor switch is used to clamp the appropriate nodes of the timing network to ground. Saturation voltage is typically below 100 mV for currents of 5 mA or less, and off-state leakage is about 20 nA (these parameters are not specified by all manufacturers, however).

Maximum collector current is internally limited by design, thereby removing restrictions on capacitor size due to peak pulse-current discharge. In certain applications, this open collector output can be used as an auxiliary output terminal, with current-sinking capability similar to the output (pin 3).

2.1.3 Basic Operating Modes

With the definition and functional description of the 555 timer just completed, we have now reached a point where the basic modes of operation of the device can be discussed in more detail. In essence, this amounts to only two modes: the monostable (or one-shot) mode and the astable (or free-running) mode. Both of these modes of operation have been discussed at length in conceptual terms; this section relates the practical operating points of the 555 to the previous material.

The two basic operating modes have a great number of variations; these are treated as specific design examples within the applications section (Part II). The 555, being such a versatile device, also has a virtually limitless number of possible operating options not necessarily directly related to the monostable and astable modes. These are much more difficult to categorize, but they will also in some way be dependent upon the internal structure just described. These more esoteric operating modes are also treated as specific design examples in the applications section.

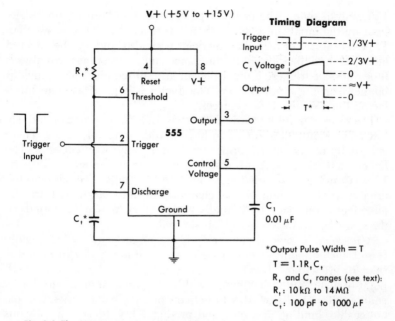

Fig. 2-4. The 555 timer connected as a triggered monostable—its most basic mode of operation.

Monostable Mode

In Fig. 2-4, the 555 is shown connected in its most basic mode of operation—as a triggered monostable. One immediate observation to be made is the utter simplicity of this circuit; it consists of only the two timing components, R_t and C_t; the timer itself; and bypass capacitor C_1 (C_1 is not absolutely essential for operation but is recommended for noise immunity).

When the trigger input terminal is held higher than $1/3V+$, the timer is in its standby state and the output is low. When a trigger pulse appears with a level less than $1/3V+$, the timer is triggered and starts its timing cycle. The output rises to a high level near $V+$; at the same time C_t begins to charge toward $V+$. When the C_t voltage crosses $2/3V+$, the timing period ends with the output falling once again to zero, ready for another input trigger. This action is graphically illustrated in the timing diagram of Fig. 2-4.

In this most simple circuit it should be noted that there are no trigger input conditioning components used. The implication of this is that the driving source in itself must be capable of satisfying the trigger voltage requirements. If the timer is operated from +5 volts in a TTL system, for instance, the input drive will automatically be TTL compatible since $1/3V+ = 1.6$ volts, which is centered in the

TTL output swing. Under this type of condition there are no restrictions on the input pulse, other than that it have a width of less than T. (Other forms of drive can be dealt with also, and will be covered later.) Due to the internal latching mechanism, the timer will always time out once triggered, regardless of any subsequent noise (such as bounce) on the input trigger. This factor is a great asset in interfacing the 555 with noisy sources.

The output pulse width is defined as $1.1R_tC_t$, and with relatively few restrictions, R_t and C_t can have a wide range of values. There is actually no theoretical upper limit on T, only practical ones. The lower limit is 10 μs. You may then consider the range of T to be 10 μs to infinity, bounded only by R and C limits. Techniques covered in a later section of this chapter will illustrate how T can be effectively multiplied by virtually any number to achieve periods of days, weeks, and even months if desired.

A reasonable lower limit for R_t is on the order of 10 kΩ, mainly from the standpoint of power economy. (Although R_t can be lower than 10 kΩ without harm, there is no need for this from the standpoint of achieving a short pulse width.) A practical minimum for C_t is about 100 pF; below this the effects of stray capacitance become noticeable, limiting accuracy and predictability. Since it is obvious that the product of these two minimums yields a T that is less than 10 μs, there is much flexibility in the selection of R_t and C_t. Usually C_t is selected first to minimize size (and expense); then R_t is chosen.

The upper limit for R_t is on the order of 13 MΩ but should be less than this if all of the accuracy of which the 555 is capable is to be achieved. The absolute upper limit of R_t is determined by the threshold current plus the discharge leakage when the operating voltage is +5 volts. For example, with a threshold plus leakage current of 120 nA, this gives a maximum value of 14 MΩ for R_t (even this value may be optimistic). Also, if the C_t leakage current is such that the sum of the threshold current and the leakage current is in excess of 120 nA, the circuit will never time out because the upper threshold voltage will not be reached. Therefore, it is good practice to select a value for R_t so that, with a voltage drop of 1/3V+ across it, the current through it will be much larger than the threshold current plus total leakage currents. This current value should be at least 10 times the threshold current plus leakage current. For best accuracy, the value should be 100 times more, if practical.

From the preceding, it should be obvious that the real limit to be placed on C_t is its leakage, not its capacitance value. In practice, however, this becomes one of capacitance value, since larger-value capacitors have higher leakages as a fact of life. Low-leakage types are available in values up to about 10 μF, however, and are preferred for long timing periods. If low-leakage units higher than this can be found, there is no limit from a circuit standpoint to using them, even up to 1000 μF.

The ultimate criterion of the components selected for R_t and C_t is the degree of accuracy desired (or expected). In general, the selection of R_t and C_t is not a trivial task because the inherent precision of the 555 is better than that of most resistors and capacitors. A detailed discussion of the selection rationale for timing components is taken up in Chapter 3, and this should be carefully studied before designs are attempted.

As previously mentioned, input trigger source conditions can exist that will necessitate some type of signal conditioning to ensure compatibility with the triggering requirements of the 555. One example of such a conditioning circuit is shown in Fig. 2-5. Here, input components C_1, R_1, and D_1 have been added, for two reasons. C_1 and R_1 form a pulse differentiator to shorten the input trigger pulse to a width less than 10 μs (in general, less than T). Their values (and relative quality) are not critical; the main criterion is that the width of the resulting differentiated pulse (after C_1) should be *less* than the desired output pulse for the period of time it is below the 1/3V+ trigger level. This effect is shown in the waveform sketch in Fig. 2-5.

The pulse as it exists at the R_1–C_1 junction will rest quiescently at the base line of V+, since R_1 is referred to V+. Therefore, the positive-going edge of the pulse would result in a voltage rise above V+, were it not for D_1. Diode D_1 is simply a switching diode connected to clamp positive excursions to the V+ level. This circuit will

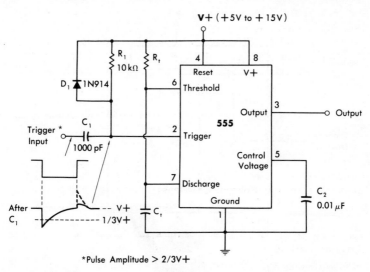

*Pulse Amplitude > 2/3V+

Fig. 2-5. The basic 555 monostable circuit with input trigger conditioning components added.

operate satisfactorily if the input pulse has the same amplitude as V+ and has a fast fall time.

Some further refinement of the input trigger circuit may be necessary if the input pulse has a peak amplitude that is less than the 555 supply voltage. For example, the circuit of Fig. 2-5 will not work when driven from a 5-volt TTL source with a timer V+ of 15 volts, since the 5-volt p-p amplitude is less than 2/3V+ (10 volts). In this type of situation, the trigger input can be biased to a level closer to the 1/3V+ threshold, thus increasing sensitivity. Fig. 2-6 illustrates this solution to the problem. Here, resistor R_2 has been added to the previously described differentiator, forming a voltage divider that will have a dc base line of 1/2V+. This biases the trigger input at this level; therefore, the amplitude of the trigger pulse need only be the difference in this dc level and 1/3V+, or simply 1/6V+. In the example mentioned, a 5-volt TTL source with a timer V+ of 15 volts, will work satisfactorily, as the 5-volt p-p amplitude is greater than 1/6 of 15 volts. The exact R_1–R_2 bias level used is not critical and may be adjusted to suit differing input requirements.

Astable Mode

The 555 connected as an astable timer is diagrammed in Fig. 2-7. This circuit also uses a minimum number of parts: the three timing components—R_{t_a}, R_{t_b}, and C_t; the timer itself; and bypass capacitor C_1. Upon startup, the voltage across C_t will be low, which causes

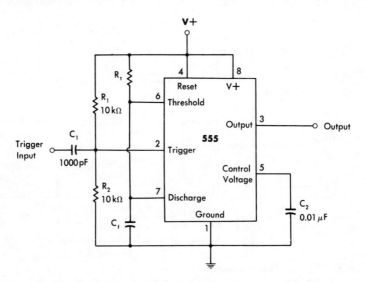

Fig. 2-6. The addition of R_2 improves the sensitivity of the input trigger conditioning circuit.

the timer to be triggered via pin 2. This forces the output high, turning off the discharge transistor and providing a current path for charging C_t via R_{t_a} and R_{t_b}. C_t charges toward V+ until the voltage reaches a level of 2/3V+, whereupon the upper threshold is reached, causing the output to go low. Capacitor C_t then discharges toward ground via R_{t_b} until its voltage reaches 1/3V+, the lower trigger point. This triggers the timer once again, beginning a new cycle. The timer then continues to oscillate between the 2/3V+ and 1/3V+ comparator threshold levels, forming a triangular timing ramp. The

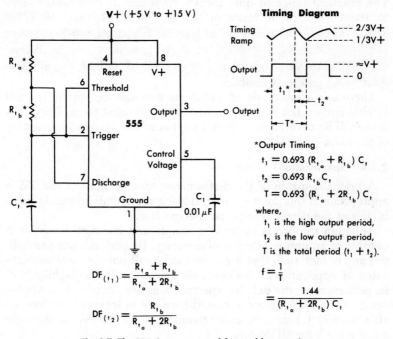

$$DF_{(t_1)} = \frac{R_{t_a} + R_{t_b}}{R_{t_a} + 2R_{t_b}}$$

$$DF_{(t_2)} = \frac{R_{t_b}}{R_{t_a} + 2R_{t_b}}$$

Fig. 2-7. The 555 timer connected for astable operation.

time duration of the high output period is t_1, and the low output period is t_2. Their sum is the total period, T. The frequency of operation is simply the reciprocal of T. The duty factor for either the high or low output state is simply that period divided by the total period.

Operating restrictions of the astable mode are few, and some are similar to monostable operation. The upper frequency limit is on the order of 100 kHz for reliable operation, due to internal storage times. There is no theoretical limit on the lower frequency, only that imposed by R_t and C_t limitations.

The limits on C_t are identical to those in the monostable mode. The maximum value of $R_{t_a} + R_{t_b}$ is the same as that for R_t of the monostable, as they are functionally equivalent. This limit is 14 MΩ or less.

Many applications may demand specific duty factors, which can be programmed (within limits) by the ratios of R_{t_a} and R_{t_b}. As R_{t_b} becomes large with respect to R_{t_a}, the duty factor approaches 50% (or square-wave operation), which can be noted from the duty factor expression. Conversely, as R_{t_a} becomes large with respect to R_{t_b}, the duty factor increases toward unity (100%) as R_{t_b} approaches zero.* R_{t_b} must not be allowed to reach zero, however. The practical range of duty factors, therefore, is from nearly 50% to about 99%; or, in terms of R_{t_a}–R_{t_b} ratios, R_{t_b} may be 1/100 of R_{t_a}. There is no limit to what fraction R_{t_a} may be of R_{t_b}, except for the absolute value restrictions. If R_{t_a} is to be a very small fraction of R_{t_b}, its value can be as low as 1 kΩ, if desired, the ultimate limit being power dissipation.

There are a multitude of variations that can be applied to this astable circuit, but it is shown here in its simplest form. The variations will be covered in a special section of the applications portion of the book (Part II).

2.1.4 Specifications

A brief discussion of the performance specifications of the 555 is appropriate at this point in order to bring the overall operating capabilities of the device into proper perspective.

Although the 555 is basically a simple and low-cost device, it is capable of an unusual degree of accuracy. Its performance capabilities are, in fact, such that it can be used in all but the most sophisticated of applications. This discussion covers only the highlights of its performance; the detailed specifications are contained in Appendix A. The specifications discussed here are in terms of commercial (0°C to +70°C) devices, as are the application discussions throughout the book for all devices.

First of all, consider the basic accuracy of the 555 in terms of the fundamental monostable timing expression, $T = 1.1R_tC_t$. There is a typical initial error of only 1% due to timer imperfections (R_t and C_t tolerance errors must be considered separately). For astable operation the error is somewhat greater, typically about 2% (this parameter is not specified by all 555 manufacturers, however).

Drift with temperature is typically only 50 ppm/°C (or 0.005%/°C) for the monostable mode. Drift in the astable mode, like the initial accuracy, is somewhat greater, or about 150 ppm/°C (not

---

* This definition of duty factor is based on the ratio of output high time (true in a logical sense) which is t_1, to total time, T.

specified by all manufacturers). These parameters apply for operation at a V+ of both +5 volts and +15 volts.

The drift with supply voltage, by design ideally zero, does have an error coefficient due to device limitations. This is typically 0.1% of timing error per volt of supply voltage change, which is still quite small.

The output-current drive capability has already been mentioned; i.e., the ability to drive both sink and source loads of up to 200 mA, and TTL compatibility when operated from 5 volts.

The input trigger and timing nodes are operable at low currents, which permits a wide range of timing resistor values. In addition, the reset function is available and it is also TTL compatible.

Finally, the device consumes a moderate amount of power, ranging from 3 mA at 5 volts to 10 mA at 15 volts (exclusive of load current).

Many of the specifications and performance features of the 555 have carried over to other similar timers; thus they are obviously well accepted. This is specifically reflected in the next device to be discussed—the 556.

2.2 THE 556 DUAL-UNIT GENERAL-PURPOSE TIMER

The NE556 timer, also first introduced by Signetics, contains two 555-type timers in the same package, with common power-supply and ground pins. It is supplied in a 14-pin dual in-line package, and each half has virtually identical electrical specifications.

2.2.1 Functional Diagram and Schematic

A functional block diagram of the 556 is shown in Fig. 2-8, and as will be noted, it is identical to the 555. Pin connections are the sole difference, since the 556 is in a 14-pin package and, being a dual unit, has duplicate pins for all but the power-supply and ground leads.

In the diagram shown, the first pin designated refers to the "A" side of the 556; the second pin refers to the "B" side. The schematic of the 556 is shown in Fig. 2-9. This also is identical to the 555 (only one-half of the dual circuit is shown). A photomicrograph of the 556 silicon chip layout is shown in Fig. 2-10.

2.2.2 Definition of Pin Functions

Without exception, all pins of each half of the 556 perform a function exactly like their 555 counterparts. Therefore, it is valid to use the descriptions given in Section 2.1.2 (the specific 555 pin functions) as applicable to the 556; the only differences will be the pin numbers.

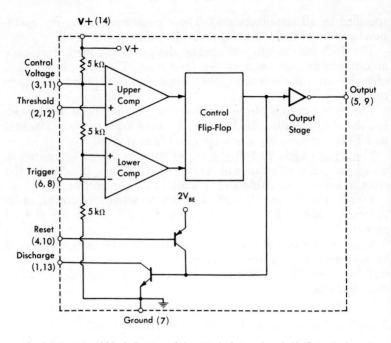

Fig. 2-8. Functional block diagram of the 556 dual timer (one-half of circuit shown).

2.2.3 Basic Operating Modes

The basic operation of the 556 in both the monostable and astable modes is identical to the 555, with the difference of duality, of course.

2.2.4 Specifications

The area of specifications is one in which there are minor differences between the 555 and 556. In general, it may be said that each half of the 556 will match or exceed the performance of a 555. With the 556, some specifications have been improved, while others have been added. Only those specifications that differ from the 555 are discussed here; the remainder are the same. For reference, a 556 data sheet is included in Appendix A.

The initial monostable timing accuracy of the 556 is specified as 0.75% typical, which is slightly better than the 555. In addition, specifications are added for the astable-mode timing accuracy. Initial accuracy is 2.25%, and drift with temperature is 150 ppm/°C. Supply-voltage sensitivity in the astable mode is typically 0.3% per volt. Threshold current in the 556 is improved over that of the 555, typically being only 30 nA (versus 100 nA in the 555).

An additional specification for the 556 is discharge leakage current, which is typically 20 nA. In practice, this current should be

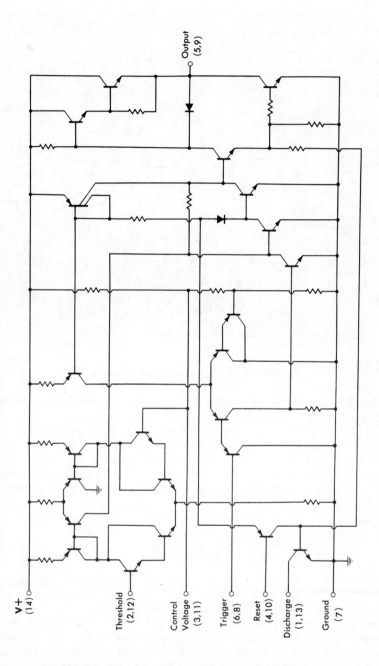

Fig. 2-9. Schematic of the 556 dual timer (one-half of circuit shown).

35

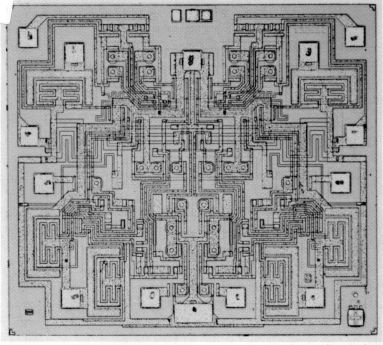

Fig. 2-10. Photomicrograph of the 556 silicon chip.

added to the threshold current to determine the total timer error current for the discharge-off period of the timing cycle.

The 556 also has specifications for matching characteristics, which define the allowable differences between sections A and B of the device. Initial timing accuracy will typically match within 0.1%, and timing drift with temperature will match within 10 ppm. Drift with supply voltage will match within 1.2%. Power drain of the 556 is double that of a 555.

2.3 THE 322 AND 3905 WIDE-RANGE, PRECISION, MONOSTABLE TIMERS

The LM322 and LM3905 timers, manufactured by National Semiconductor, were the first timers introduced that departed to any great extent from the established 555 design concept. These units provide improved performance over the 555 in a number of areas: wider supply-voltage range, greater supply immunity, a more flexible and higher voltage output stage, a wider range of timing period capability, and greater timing accuracy. One feature that was sacrificed, however, was that of astable operation, as both the 322 and 3905 are basically monostable timers. Thus, although both the 322 and 3905 possess the technical performance that justifies the term *precision timers,* they are not completely general purpose in the same sense as the 555 and 556 devices are.

The 322 is available in either a 10-pin TO-5 style metal-can package, or in a 14-pin dual in-line plastic package (DIP), while the 3905 is available in an 8-pin dual in-line (MINIDIP) plastic package.

2.3.1 Functional Diagram and Schematic

A functional block diagram of the 322/3905 is shown in Fig. 2-11. In some senses, this is similar to the 555 timer, while in others it is different. In this diagram, the first designated pin number is for the 322 (in the 14-pin DIP), while the second number pertains to the 3905.

The part of the diagram that is similar to the 555 is the arrangement of the comparator, the flip-flop, the discharge transistor (Q_1), and the threshold reference divider (R_{16}–R_{17}). In the 322 and 3905, the timing voltage reference is made 0.632 times the timing voltage, in order to make the timing equation equal to $T = R_tC_t$.

An important difference from the 555 is the fact that the timing voltage in the 322/3905 comes from an on-chip voltage regulator, which supplies a constant 3.15 volts to the divider. The voltage regulator is also made available externally for the connection of a timing resistor. Although the basic timing scheme of the 555 (i.e., charging from and comparing against a fraction of the supply voltage) is in theory insensitive to supply changes, it does lose some accuracy if the supply voltage changes during the timing interval. The 322/3905 design eliminates supply voltage fluctuations as a source of error by virtue of the 3.15-volt regulator. The comparison voltage in the 322/3905 is nominally 2 volts (0.632 times 3.15 V), established by R_{16}–R_{17}. In the 322, this point is made externally available; in the 8-pin 3905, it is not.

The R/C node is the junction point of the external timing resistor and capacitor, and the discharge transistor is connected internally to this point. Threshold current is very low in these devices, on the order of 300 pA for both the 322 and the 3905. In the 322, there is an optionally used terminal, the boost terminal (pin 11), which allows higher comparator speed. With the boost terminal connected externally to V+, the threshold current of the 322 becomes 30 nA.

Triggering in the 322 and 3905 is accomplished by a positive pulse at the trigger input. This pulse sets the latch, starting the timing cycle. However, a difference in these timers over those types previously described is the unique output stage used. The signal from the

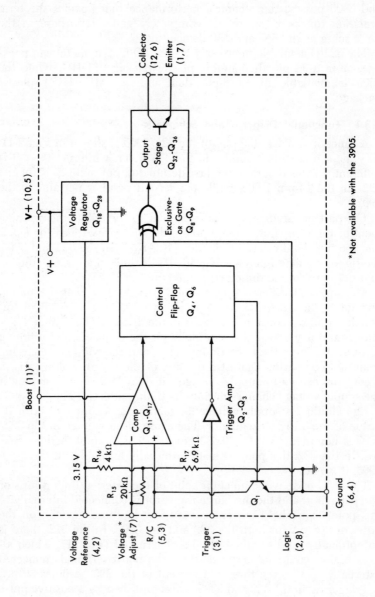

Fig. 2-11. Functional block diagram of the 322/3905 precision monostable timers.

38

latch does not simply yield a high output, as in 555-type timers, for two reasons. The first of these is an exclusive-OR gate, which has as its two inputs the latch output and the logic input pin. The function of this gate is to control the relative on/off state of the output stage by the logic input. When the logic pin is low, the output stage is "on" during the timing cycle and is "off" otherwise. When the logic pin is high, the output stage is "off" during the timing cycle and is "on" otherwise.

The output stage is equivalent to a floating npn power transistor, and both collector and emitter pins are brought outside the package for external connection. Thus, the output stage can be wired in either a common-collector or a common-emitter fashion. This composite transistor has a voltage standoff capability of 40 volts, and it is also current limited internally. With the flexibility that this output stage and the gating combination possess, many different functions can be performed.

The schematic of Fig. 2-12 illustrates the internal circuitry of the 322/3905. Again, the first designated pin number refers to the 322 (in the 14-pin DIP), while the second number refers to the 3905. The trigger input amplifier is made up of Q_2 and Q_3, which in turn drive the npn/pnp latch, Q_4 and Q_6. The discharge transistor, Q_1, is driven from the latch and Q_2.

Transistors Q_{14}–Q_{17} form a pnp input differential comparator, and Q_{11}–Q_{13} further amplify the output of Q_{15}–Q_{16}. The comparator is interfaced to the latch by Q_7. It also drives the exclusive-OR gate, Q_8–Q_9. The boost terminal of the 322 is connected to the emitter of Q_{30} within the comparator bias circuitry. Jumpering this pin to V+ turns on Q_{30}, which increases the operating current of Q_{14}–Q_{17}, and thus the comparator speed.

Transistors Q_{18}–Q_{28} comprise the voltage regulator. The actual regulator is Q_{18}–Q_{24}; Q_{25} and Q_{26} serve as start-up components only. The output voltage is 3.15 volts, and light external loads may be connected.

The output stage, consisting of Q_{32}–Q_{36}, has a "floating" base drive supplied by Q_{31}. This drive is either passed to the output or shunted away by Q_8 (the exclusive-OR gate) to control the conduction state of the output. Within the output stage, Q_{32} provides current limiting at 120 mA, and Q_{34}–Q_{35} serve as voltage clamps to prevent excess storage time in Q_{36}. A photomicrograph of the LM322 silicon chip is shown in Fig. 2-13.

2.3.2 Definition of Pin Functions

Further insight into the operation of the 322/3905 devices may be gained by discussing each pin. Once again, the first designated pin number refers to the 322 (in the 14-pin DIP), while the second num-

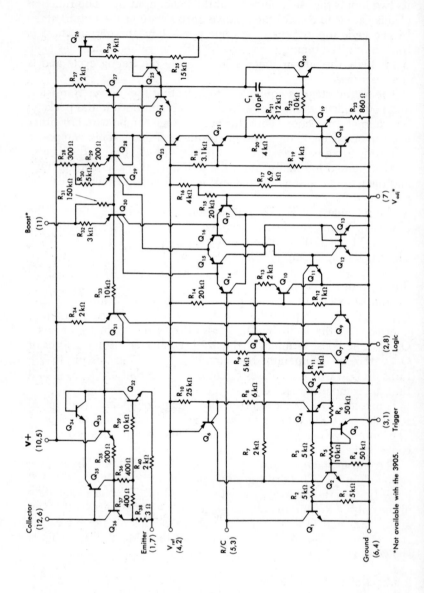

Fig. 2-12. Schematic of the 322/3905 precision monostable timers.

ber refers to the 3905. The reader is referred to the 322/3905 data sheets in Appendix A.

V+ (Pins 10, 5)

The V+ pin is the most positive supply terminal of the 322/3905. Operation is highly independent of the voltage applied. Supply current is typically 2.5 mA and is independent of the supply voltage (this will be higher if the reference is loaded). The current drive capability of the output stage is also independent of the supply voltage.

Ground (Pins 6, 4)

This pin is the most negative supply potential of the device (normally connected to circuit common when operating from positive supply voltages).

Collector (Pins 12, 6)

This output pin is the collector of the floating npn output transistor. This transistor has a minimum breakdown of 40 volts and is current-limited to 120 mA. For outputs taken from the collector, the emitter is normally grounded. The output is taken from the collector with a load referred to V+, or other positive voltage of 40 volts or less.

Emitter (Pins 1, 7)

This output pin is the emitter of the npn output transistor. For outputs taken from the emitter, the collector is tied to V+ (or other positive voltage). The output is taken from the emitter with a load referred to ground. The collector may be connected to a voltage higher than V+, but the emitter will pull up (when "on") to a voltage somewhat less than V+.

Voltage Reference (Pins 4, 2)

This pin is the output of the internal 3.15-volt regulator. Loads of up to 5 mA may be applied, if desired. In normal use, the timing resistor is connected from this pin to the R/C pin. Drift of the reference is typically 0.01%/°C, making this reference voltage quite useful in external circuitry.

Voltage Adjust (Pin 7 on 322; Not Available on 3905)

This pin allows access to the 2-volt comparator reference point. Use of the pin is optional; it may be used to trim the timing period, if desired, or to prematurely end the timing cycle with a negative-going voltage. When this pin is not used, noise immunity will be enhanced if it is bypassed with a capacitor of from 0.01 μF to 0.1 μF.

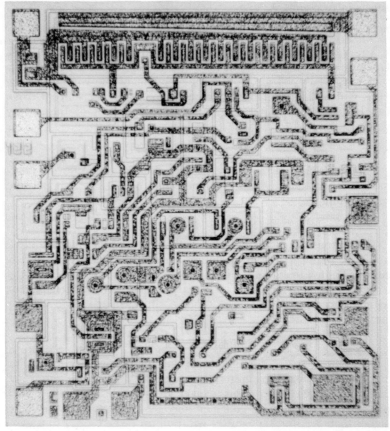

Fig. 2-13. Photomicrograph of the LM322 silicon chip.

The voltage-adjust pin, in conjunction with the R/C pin, allows access to both inputs to the comparator. This permits use of the device as a general-purpose, low-input-current comparator with a common-mode input range of zero up to 5 volts. External voltages applied to the voltage-adjust pin should be between zero and +5 volts for safe operation.

R/C (Pins 5, 3)

This pin is the node where the timing resistor and capacitor are normally connected. Internally, the R/C pin is connected to the comparator input and the collector of the discharge transistor (Q_1). The timing threshold voltage is +2 volts. When the voltage across the timing capacitor reaches +2 volts, the comparator changes states, ending the timing cycle.

The threshold current is typically 300 pA for the 322 operating in the unboosted mode, and for the 3905. When the 322 is operating in the boosted mode, the threshold current increases to 30 nA. For safe operation, external voltages applied to the R/C input should be between zero and +5 volts.

Trigger (Pins 3, 1)

This pin is used to start the timing cycle with a positive-going pulse. The trigger threshold is TTL compatible, with a typical threshold voltage of +1.6 volts. Current at threshold is 20 μA, and the input is overvoltage protected for voltages up to ±40 volts.

The timer will not retrigger if the trigger cycle is held high during the timing cycle but will time out. However, the timing capacitor will not be discharged until the trigger input is lowered below the threshold (this does not affect the output).

Logic (Pins 2, 8)

This pin determines the state of conduction of the output transistor during the timing cycle. When the logic pin is high, the output transistor is "off" during the timing period ("on" otherwise). When the logic pin is low, the output transistor is "on" during the timing period ("off" otherwise).

The logic input switching threshold is 150 mV, and 150 μA of current must be sunk by the source in the low state. Safe voltage applied to this pin is zero to +5 volts.

Boost (Pin 11 on 322; Not Available on 3905)

This pin increases the speed of the comparator when connected to the V+ pin. It is used when operating at short timing periods ($\leq$1.0 ms) for greater accuracy.

2.3.3 Basic Operation

The basic mode of operation for the 322/3905 is that of a monostable timer. Astable operation can be achieved with some additional circuitry, if desired. This section deals with the practical points of applying the device in its basic mode.

Monostable Mode

There are two different ways that the 322/3905 timer can be used in the monostable mode of operation: with the output taken from the collector, or with the output taken from the emitter. Fig. 2-14 illustrates the collector-output option, where the emitter of the output transistor is connected to common (ground). R_L is the load re-

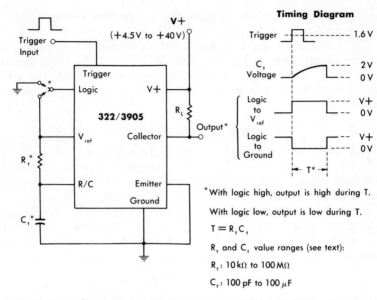

Fig. 2-14. The 322/3905 connected in the basic monostable mode of operation, with the output taken from the collector terminal.

sistor, and is referred here to V+. R_t and C_t are the timing components, and the logic pin may be connected to either V_{ref} or ground (high or low), depending on the desired output state.

When the input trigger exceeds +1.6 volts, the timer fires and begins a timing cycle. Capacitor C_t charges toward V_{ref} and, assuming the logic pin to be high, the output goes high during the timing cycle. When the C_t voltage crosses +2 volts (0.632 times V_{ref}), the timing cycle ends, C_t discharges, and the output returns to its low state. Had the logic pin been wired low, the timing period would be identical but the output state would be reversed (see the timing diagram in Fig. 2-14). This collector-output option is useful when current-sinking loads, such as TTL logic, are to be driven.

Fig. 2-15 shows the emitter-output option. This is similar to the collector-output option, but the load resistor, R_L, is placed between the emitter and ground, and the collector is wired to V+. With the logic input high, the output of this circuit is low during the timing cycle. If the logic input is low, the output is high during the timing cycle. The high-state output will typically pull up to within 1.8 to 2 volts of V+ in this circuit, depending on the load current. This circuit is useful for current-source-type loads.

Limitations on timing and timing components are identical for both circuits, and are generally few. The 322 (with the boost option) is useful down to a few microseconds in pulse width. At the other

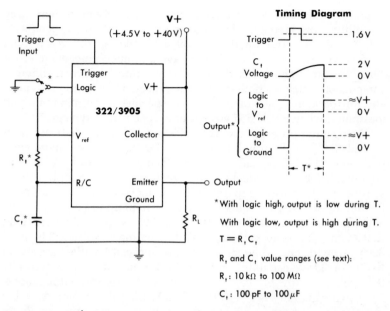

Fig. 2-15. The 322/3905 connected in the basic monostable mode of operation, with the output taken from the emitter terminal.

end of the timing range, an unboosted 322 (or 3905) can achieve timing periods of hours or more, limited only by the timing components.

Due to the extremely low comparator current capability of the 322/3905, the allowable range of R_t is from 10 kΩ to 100 MΩ. The ultimate limit is more a function of the availability of high-value resistors. Capacitor C_t can range from 100 pF on the low end (a practical limit due to stray capacitance) up to 100 μF on the high end. The real limit on C_t will also be one of availability. With the 322 and 3905 timers, the operating range is essentially unrestricted from the standpoint of the device itself; the limitations are practical ones set by R_t and C_t.

Astable Mode

Because these devices do not offer the capability of astable operation by themselves, this mode will not be treated in this section but will be covered in the applications section of the book.

2.3.4 Specifications

A discussion of the specifications of the 322 and 3905 will serve to bring their performance capability into overall focus. Of all the devices discussed thus far, these two are capable of the highest degree

45

of performance and best overall accuracy, in addition to their basic flexibility. For reference, data sheets are included in Appendix A.

Initial timing error for the 322/3905 is specified differently than for 555-type devices, which simply have a lumped percentage error. The 322/3905 units break the error down into its several different components, such as timing-ratio tolerances, C_t saturation voltage, etc. Furthermore, some of the error is specified for a minimum and maximum, whereas the 555 error is only specified as typical.

Timing ratio error for the 322/3905 is listed as a worst-case limit of ±3.2%. Normally it will be much better than this and more comparable to the 1% typical of the 555. Temperature drift is listed as 30 ppm/°C typical, which is an excellent figure. Supply-voltage sensitivity is not listed separately but is included in the total error for timing ratio. By itself it would be about 50 ppm/V.

Output drive capability of the 322/3905 is quite different than that of the 555 devices, because of the floating transistor. The 322/3905 can withstand voltages of 40 volts and can handle currents up to 50 mA. Drive capability in the collector-output mode is typically 0.25 volt at 8 mA; 0.7 volt at 50 mA. In the emitter-output mode, the drive capability is 1.8 volts at 3 mA; 2.1 volts at 50 mA. A big point in favor of the uncommitted output arrangement is the ability to drive loads that are referred to voltages higher than V+, up to 40 volts.

Trigger voltage is typically 1.6 volts which is TTL compatible and requires 25 μA of source current. Minimum trigger pulse width is 0.25 μs.

A significant difference between 555 devices and the 322/3905 is their very low input threshold currents. For the 3905 (and the 322 unboosted) this is only 300 pA typical. For the 322 in the boosted mode, the threshold current rises to 30 nA.

Another major point of usefulness for the 322/3905 is the 3.15-volt reference output. Although this voltage has no influence on the timing accuracy, it has a specified tolerance of ±5% and good regulation. This feature can be a valuable asset in system design.

2.4 THE 2240, 2250, AND 8260 PROGRAMMABLE TIMER/COUNTERS

The 2240, 2250, and 8260 are three different types of devices which fall into a special class of timer ICs—that of programmable timer/counters. These units include a timing section made up of a 555-type oscillator, followed by a counter section. The timing period of the counter is externally programmable by the user.

In simplest block diagram terms, the timer portion generates a basic timing pulse of a period, T. This is then multiplied (or counted) by the counter to effectively increase the timing period by a desired multiplication factor. The multiplication factor can be made externally variable, thus the term *programmable timer/counter*. Differences between the various devices mentioned are generally in terms of the functioning of the counter section, the timer sections all being similar.

The XR-2240 is manufactured by Exar Integrated Systems, who pioneered the concept of the timer/counter. It is a binary programmable device with an 8-bit counter section. Timing is programmable from $1R_tC_t$ to $255\ R_tC_t$, where R_t and C_t are the timing components that define the basic timing interval, which is $T = R_tC_t$. Two or more 2240 devices can be cascaded to extend the timing interval indefinitely, counting in binary fashion.

The XR-2250, also manufactured by Exar, is a bcd (binary-coded decimal) programmable device. The time-base section of the 2250 is similar to the 2240, but the counter is an 8-bit design arranged to count in decimal fashion over two decades. It is programmable from $1R_tC_t$ to $99R_tC_t$. Two or more 2250 devices can be cascaded to extend the timing interval indefinitely, counting in decimal fashion.

The ICL 8260, manufactured by Intersil, is designed for seconds/minutes/hours counting applications. The time-base section is similar to the 2240 and 2250, and the counter section is programmable from 1 to $59R_tC_t$. Two 8260s can be cascaded to count seconds and minutes.

This programmable timer/counter section is somewhat larger in scope than the previous sections of this chapter because it discusses an entire family of timing devices, each of which is a large-scale system in itself. Although there are differences between the timers covered, these are differences of detail rather than concept. Therefore, the 2240 will be discussed initially in detail, and when its operation is understood, it will be relatively easy to transfer your thinking to the 2250 and 8260 types.

2.4.1 The 2240 Binary Programmable Timer/Counter

The XR-2240 is produced by Exar, as mentioned previously, and it is second-sourced by others under the 2240 part-number designation. It is also second-sourced by Intersil under the 8240 designation. The 2240 can be operated in either the basic monostable or astable timing modes, as well as a variety of other modes. It is packaged in a 16-pin dual in-line package.

Functional Diagram

A functional block diagram of the 2240 is shown in Fig. 2-16A. The two main functional portions of this diagram are the time-base section and the counter. The time-base section may be seen at the left

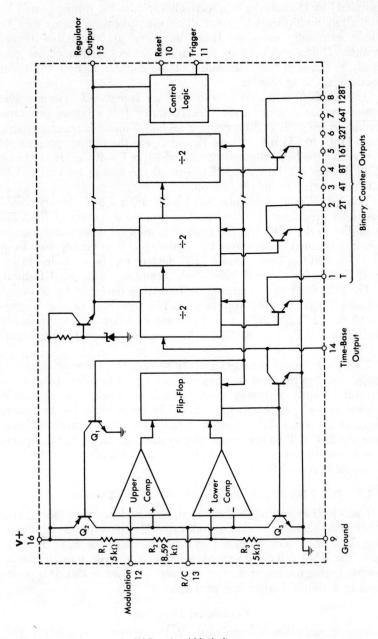

(A) Functional block diagram.

Fig. 2-16. The 2240 binary

of the diagram and is made up of a 555-type astable oscillator, with a buffered output at pin 14. The counter is seen at the right and consists of 8 binary stages, with a buffered output available from each. Each output is low for the multiple of the time-base period shown. The T output (pin 1), for instance, is low for a period, T, while the 2T output (pin 2) is low for a period, 2T, etc.

A third subsection is the control logic, a circuit consisting of a latch that is set and reset by pins 11 and 10, respectively. This circuit controls the timer/counter, resetting all counter stages when commanded, and starting the timer circuit upon command by a trigger.

The control functions and counter stages are powered by an on-chip voltage regulator, which produces a nominal 4- to 6-volt output. This is made available at pin 15. The time-base portion of the circuit is connected directly across V+ and uses the 555 scheme of charging from and comparing against the same (V+) voltage, for timing-period supply-voltage independence.

In operation, an external timing resistor and capacitor are connected to the R/C node, somewhat similarly to the connection for the 555 monostable. The capacitor is charged toward V+ via R_t, and the timing ramp is compared to a fraction of V+ generated by the R_1–R_2–R_3 divider. The upper threshold is 0.731 of V+. Capacitor C_t is discharged toward ground by the discharge transistor, and the lower threshold is 0.269 of V+.

Although on the surface this scheme appears similar to the 555 astable timing mode, there are some important differences. First is

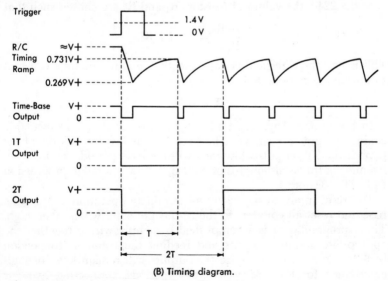

(B) Timing diagram.

programmable timer/counter.

the fact that this circuit is a *triggered* astable, and will always operate in an astable mode, even if the timer/counter as a whole is used as a monostable. An understanding of how the circuit functions may be gained from the timing diagram of Fig. 2-16B.

In the reset or standby state, the time-base oscillator is inhibited by Q_2, which clamps the R/C node to nearly the V+ level. This may be noted on the timing diagram in the sketch for the timing ramp. When the circuit is triggered, Q_2 turns off, removing the clamp. The discharge transistor, Q_3, then rapidly brings the capacitor voltage down to the lower comparator threshold. This changes the state of the oscillator flip-flop, and C_t now charges toward V+. When the upper threshold is reached, Q_3 discharges C_t again. The oscillator will then continue to cycle until the control logic is reset.

Although the time-base operation is astable, the timing is highly asymmetrical because Q_3 discharges C_t very rapidly but is charged slowly through R_t. It is for this reason that the timing expression for T is stated simply as a single period—the discharge time is short enough to be neglected. This discharge time period generates the short negative-going time-base output pulses. Note that the very first discharge cycle is from V+, rather than 0.731 of V+. This factor will result in only a slightly longer first timing period than the rest because the discharge current for C_t is high. The timing scale is exaggerated in the diagram to illustrate this point. In practice, the error due to this effect is minimal, particularly for multiple time counts.

In the 2240, the values of resistors R_1 and R_2 are chosen such that

$$\frac{R_1 + R_2}{R_1} = \epsilon = 2.7183.$$

This makes the threshold voltage exactly equal to one time constant so the timing period, T, is then simply:

$$T = R_t C_t.$$

In the timing diagram of Fig. 2-16B, two of the eight available counter outputs are shown. The 1T output is low for one time-base period; the 2T output is low for two time-base periods; and the remaining outputs, although not shown, follow a similar progression (4T, 8T, 16T, etc.).

The reset input on the 2240 will terminate operation of the time base and reset all counters to the zero-count state when taken high. Timer programming is accomplished by simply wiring together the appropriate counter outputs and feeding the common connection back to the reset input. Since the outputs will remain low (or logically true) for their respective time periods, connecting two (or more) together causes the common low output period to be the *sum* of the individual timing periods. Thus, a period of 3T would be achieved by wiring together outputs 1T and 2T; a period of 7T would be achieved by wiring together outputs 1T, 2T, and 4T; etc. This is shown in Fig. 2-17, with an example of a programmed count of 10T (2T + 8T). Regardless of the count programmed, when the common output line goes high, the 2240 is reset, and the timing cycle is ended.

Definition of Pin Functions

This section further defines the performance of the 2240 by describing the function of each pin. The reader is referred to the 2240 data sheet in Appendix A.

V+ (Pin 16). The positive supply voltage terminal of the device. Supply range is 4 volts (minimum) to 15 volts (maximum), and the 2240 may be operated anywhere in this range without major changes in performance.

Ground (Pin 9). The most negative supply terminal of the device, usually connected to circuit common. Wiring to this pin should be of a low impedance, as it carries currents common to both the analog and digital sections.

Counter Outputs (Pins 1–8). Buffered, open-collector, npn transistor stages, one for each timing terminal on the 8-bit binary counter. Each output is capable of sinking 4 mA of current at a low-state voltage of 0.4 volt, making them TTL compatible. High, or off-state, voltage can be as high as 15 volts.

In the reset state of the device, all counter outputs will be in the high state. The true, or active, logic state for each counter output is the low state, and timing is in accordance with Fig. 2-16B. The counter outputs may be either used individually or wired together for various programmed times. The combined or "wired-or" output will remain low as long as any one of the individual outputs is low.

Trigger (Pin 11). The pin that actuates the timing cycle by setting the internal control logic, which enables the time-base oscillator. Once triggered, the device will time out and will be immune to further triggers during the timing cycle. The trigger threshold is +1.4 volts, and triggering is accomplished with a positive-going pulse. Minimum trigger pulse width is 2 μs.

When power is first applied, the device will automatically revert to the reset state. If both trigger and reset pulses are applied simultaneously, the trigger pulse overrides the reset pulse. Minimum retrigger time after a reset time is a function of supply voltage and timing capacitance (see the data sheet in Appendix A).

Reset (Pin 10). The pin that terminates the timing cycle by resetting the internal control logic. This returns the counters to their preset of zero count and disables the time-base oscillator. The threshold of the reset input is similar to the trigger input; it responds to positive-going pulses at a level of 1.4 volts.

R/C (Pin 13). The connection for the external timing resistor and capacitor. Internally, this pin is connected to the inputs of both comparators and the discharge and clamp transistors.

The time-base oscillator is astable in nature, and the waveform appearing across C_t will be an exponentially rising sawtooth with a short retrace. Timing voltage thresholds are 0.731 of V+ for the upper comparator and 0.269 of V+ for the lower comparator.

The timing sequence of time-base operation is illustrated in Fig. 2-16B. The time-base output is low (or true) when the timing capacitor is being discharged. This time is a small percentage of the total timing period. The timing expression for the 2240 is $T = R_t C_t$, as the upper comparator threshold voltage is set to a level of 0.632 of the charging voltage.

Comparator threshold currents are not specified for the 2240; this point is taken care of by specifying a maximum range of timing components (these are discussed under "Basic Operating Modes").

Modulation (Pin 12). The pin that allows access to the upper comparator reference voltage, which is normally 0.731 of V+. The dc voltages applied to this point may be used to control the time-base period, if desired. Also, the time-base oscillator may be synchronized to an external source by applying signals to this pin. Safe voltage limits for this pin are between zero and V+.

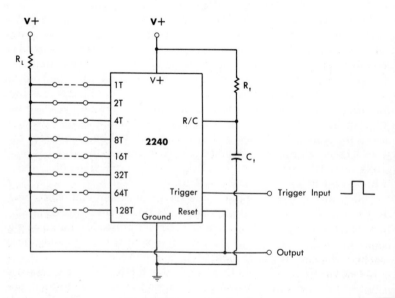

Fig. 2-17. Programming the 2240.

Time-Base Output (***Pin 14***). The output from the time-base oscillator. The pulse on this pin is high during the reset state and goes low upon triggering (and with each subsequent time-base period). Timing is as shown in Fig. 2-16B.

The time-base output is internally connected to the counter input, as shown in Fig. 2-16A. The negative-going time-base pulses trigger the counter section. An external pull-up resistor from this pin to the regulator output (pin 15) is required for operation. A value for this resistor in the vicinity of 20 kΩ is recommended, but it is not critical.

The 2240 may also be operated with an external clock source, if desired. In such a case, the time-base output pin serves as an input to the counter section. Counter input trigger threshold is 1.5 volts. Also, the counter may be disabled, if necessary, by clamping pin 14 to ground externally.

Regulator Output (***Pin 15***). The output of an on-chip voltage regulator, which powers the counter and logic sections of the 2240. If two or more 2240s are cascaded, the regulator output can be used to power the counter sections of succeeding devices without their time bases, in the interest of low power consumption. Output current from this pin should be 10 mA or less.

With a V+ of 15 volts, the regulator output will be typically 6.3 volts; it is typically 4.4 volts with a V+ of 5 volts. If the time base is to be used with a V+ of 4.5 volts or less, it is recommended that pin 16 be jumpered to pin 15.

Basic Operating Modes

The 2240 is capable of operation in either the monostable or the astable timing mode. Timing is programmable in either mode, and operation is achieved with a low number of external components.

Monostable Mode. The 2240 is shown connected for monostable operation in Fig. 2-18. In this circuit, R_t and C_t set up the time base for the desired basic period, T. Programming is accomplished by jumpers, switches, or other suitable means. The timer output appears across the common load resistor, R_L, and the output pulse width, T_o, is equal to nR_tC_t, where n is the number selected by the program switches.

As shown in the timing diagram, the output is high at a level of V+ prior to triggering. With the arrival of the trigger pulse, the output falls to a low level and the timing cycle is initiated. The time-base oscillator will continue to run until the counter reaches the count programmed by the selector switches. When this count is reached, the output rises from the low level to V+. This rise in level is fed to the reset input, which stops the time base and resets the counter. The timer is now back to its standby state, awaiting the next trigger.

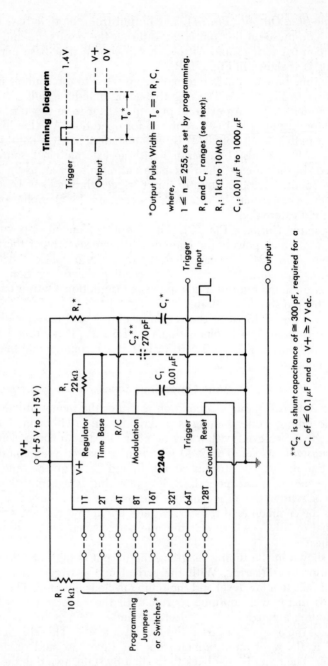

Timing Diagram

Trigger — — — 1.4V

Output — — — V+
— — — 0V

T_o^*

* Output Pulse Width $= T_o = n\,R_t\,C_t$.

where,

$1 \leq n \leq 255$, as set by programming.

R_t and C_t ranges (see text):

R_t: 1 kΩ to 10 MΩ

C_t: 0.01 μF to 1000 μF

V+ (+5V to +15V)

R_t^*

C_2^{**} 270 pF

C_1 0.01 μF

C_t^*

R/C

Regulator

Time Base

Modulation

Trigger

Reset

V+

2240

1T 2T 4T 8T 16T 32T 64T 128T

Ground

R_1 22 kΩ

R_L 10 kΩ

Programming Jumpers or Switches*

Trigger Input

Output

** C_2 is a shunt capacitance of $\cong$ 300 pF, required for a C_t of $\leq$ 0.1 μF and a V+ $\geq$ 7 Vdc.

Fig. 2-18. The 2240 programmable timer/counter connected for monostable operation.

Other circuit components are R_1, which serves as a load resistor for the time-base output, and C_t, a noise bypass to ensure noise immunity within the time-base upper comparator. The value shown for R_L is not a critical one, and the voltage to which it is returned need not be the same voltage as the chip supply, as long as it is between 5 volts and 15 volts. Thus, a 2240 may readily drive various forms of logic, as well as discrete stages.

The range of timing components useful with the 2240 is broad. It is recommended that the timing resistance be between 1 kΩ and 10 MΩ. The capacitance should be between 0.01 μF and 1000 μF. In addition to these limits, there is a maximum limit on the time-base oscillation frequency of about 100 kHz, which effectively places a minimum on T of 10 μs. All of these constraints should be met in an optimum design.

Astable Mode. The astable mode of operation for the 2240 is shown in Fig. 2-19. This circuit is similar to the monostable mode of operation, with the exception that the reset input is not connected to the output. This allows the 2240 to continue oscillation, once started by a trigger pulse. With a single counter output connected to R_L and the output bus, the 2240 will oscillate at a frequency, f_o, which is:

$$f_o = \frac{1}{2nR_tC_t},$$

where n is the count of the particular tap selected (1, 2, 4, 8, etc.). The factor of 2n is derived from the fact that, in terms of frequency, the basic timing taps of the 2240 counter are multiples of one-half period.

The circuit will not self-start with the application of power because the internal logic will revert to the reset state. A pulse applied to the trigger input will start synchronous oscillations, with timing as shown. If desired, oscillation may be halted by the application of a reset pulse, which will cause the output to go high. The circuit will then remain in that state until triggered again. If automatic power-up oscillation is desired, the trigger input can be wired to the regulator output (pin 15) which will self-start the circuit. Timing component considerations are identical to those of the basic monostable circuit.

Specifications

The specifications of the 2240 are by nature different from the previous timers discussed, since it is a much more complex device. Still there are many similarities, so the following discussion will emphasize the differences from 555-type timers. For reference purposes, a 2240 data sheet is included in Appendix A.

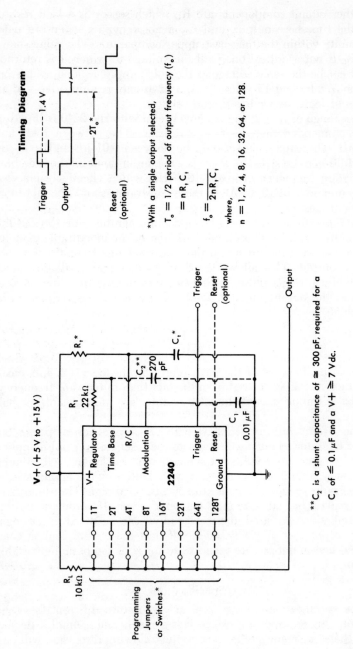

Timing Diagram

Trigger

Output ---- 1.4 V

$2T_o$ *

Reset
(optional)

*With a single output selected,

$T_o = 1/2$ period of output frequency (f_o)

$\equiv n R_1 C_1$

$$f_o = \frac{1}{2n R_1 C_1}$$

where,

n = 1, 2, 4, 8, 16, 32, 64, or 128.

V+ (+5V to +15V)

R_1*

R_1 22 kΩ

C_2** 270 pF

C_1*

2240

1T	V+ Regulator
2T	Time Base
4T	R/C
8T	Modulation
16T	
32T	
64T	Trigger
128T	Reset
	Ground

Trigger

Reset
(optional)

C_1 0.01 µF

Output

R_L 10 kΩ

Programming Jumpers or Switches *

**C_2 is a shunt capacitance of ≅ 300 pF, required for a C_1 of ≤ 0.1 µF and a V+ ≥ 7 Vdc.

Fig. 2-19. The 2240 programmable timer/counter connected for astable operation.

The 2240 operates from a similar range of power-supply voltages—+4 volts to +15 volts. Typical supply current is 4 mA at +5 volts; 13 mA at +15 volts. A substantial portion of this current is consumed by the time base; only 1.5 mA is consumed by the counter.

Timing accuracy is very good, typically 0.5% at a V+ of 5 volts. Drift with temperature varies somewhat with supply voltage, being 200 ppm/°C at +5 volts, but 80 ppm/°C at 15 volts. Supply-voltage timing sensitivity is listed as 0.08% per volt (typical) for supply voltages of 8 volts or more.

The maximum operating frequency is typically 130 kHz, as defined by the minimum timing components. It is recommended that the timing resistance be between 1 kΩ and 10 MΩ, while the timing capacitance should be between 0.01 μF and 1000 μF.

The trigger and reset inputs of the 2240 have very similar sensitivities. The voltage thresholds are 1.4 volts, the impedance is 25 kΩ, and the current is 10 μA at an input voltage of 2 volts. Trigger response time is 1 μs, while reset response time is 0.8 μs.

The counter section has a maximum toggle rate of 1.5 MHz typical. While this is above the maximum frequency of the internal time base, it does indicate that the 2240 is useful with an external clock at higher rates.

The counter trigger input (time-base output) has a trigger threshold of 1.4 volts and an impedance of 20 kΩ. For the counter outputs, rise and fall times are 180 ns with a load of 3 kΩ and 10 pF. Each output can sink 4 mA at a voltage of 0.4 volt or less. Off-state leakage is typically 0.01 μA at a voltage of 15 volts.

2.4.2 The 2250 BCD Programmable Timer/Counter

The XR-2250 is also manufactured by Exar Integrated Systems. It is second-sourced by Intersil under the 8250 designation, with a slightly different pin configuration. The 2250, like the 2240, may be operated in either the basic monostable or the astable timing mode, as well as a variety of other modes. It is packaged in a 16-pin dual in-line package.

Functional Diagram

A functional block diagram of the 2250 is shown in Fig. 2-20. This diagram is essentially the same as that for the 2240, the primary difference being in the counter section. Minor differences include a new pin designation for the modulation input (pin 15), the deletion of the regulator output, an internal load resistor on the time-base output, and the addition of a carry-out terminal (pin 12) used to cascade two or more 2250s.

In the counter section, it may be noted that there are still eight stages, with a similar arrangement of output pins. In the 2250, how-

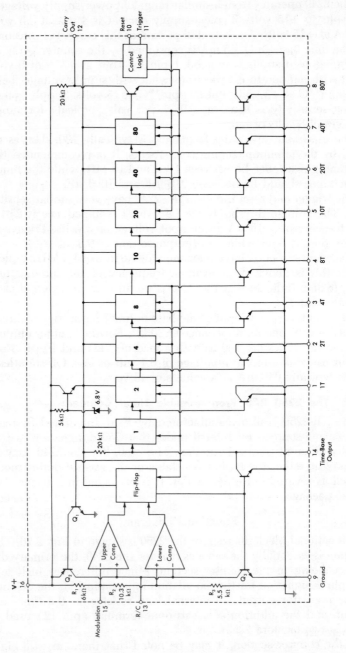

Fig. 2-20. Functional block diagram of the 2250 bcd programmable timer/counter.

ever, the counters are arranged in groups of four bits. The first four bits (1, 2, 4, and 8) are connected in a feedback arrangement so that they count in bcd from 0 to 9. The output from this first four-bit set, which may be called a *units* decade, drives a second four-bit set: 10, 20, 40, and 80.

The second set of counters is the *tens* decade, and similarly counts from 0 to 9 by virtue of feedback. As the input drive is in units of ten, however, the actual output count is in tens; therefore, this section counts from 0 to 90.

The carry-out terminal, pin 12, is used in applications where two or more 2250s are to be cascaded. This pin goes low at a count of 100 and is used to drive the time-base (counter) input of a succeeding 2250, which will then count hundreds (and thousands).

The remaining functions of the 2250 are the same as has been described previously for the 2240. Programming is accomplished by connecting the appropriate output pins to the output bus. The only difference is that the maximum count for the 2250 is 99 rather than 255 in the case of the 2240. The control logic, trigger and reset inputs, and the time-base section operate in a manner very similar to the 2240.

Definition of Pin Functions

The only additional pin on the 2250 that is functionally different from the 2240 is the carry-out pin, which is described next. There are, however, some pinout differences between the Exar 2250 and the Intersil 8250, which are noted. The reader is referred to the 2250 and the 8250 data sheets in Appendix A.

Carry Out (Pin 12 on the 2250; Pin 15 on the 8250). This pin is used to signal an overflow of the counter, corresponding to a count of 100. This output goes low at a count of 100 and is used to drive the counter of a succeeding 2250 via the time-base output (counter input) pin. The output is TTL compatible and can thus be used to drive other forms of logic if desired.

Modulation (Pin 15 on the 2250; Pin 12 on the 8250). Functionally, there is no difference between this pin on the 2240 and the 2250. In the 2250 it is placed on pin 15, whereas the 8250 retains the use of pin 12 for this function.

Time-Base Output (Pin 14). There is an internal load resistor added on this pin; therefore, a load resistor is not needed externally as is the case with the 2240.

Basic Operating Modes

There are no functional differences in the basic operating modes between the 2250 and the 2240, with the exception of the counting, which is obviously different. The basic monostable connection is similar to Fig. 2-18 (with the exception of the pinout differences), and the basic timing expression is

$$T_o = nR_tC_t.$$

In this case, however, n is variable from 1 to 99.

The basic astable connection is similar to Fig. 2-19, and the frequency expression is

$$f_o = \frac{1}{2nR_tC_t} .$$

In this case, however, n is either 1, 2, 4, 8, 10, 20, 40, or 80.

Specifications

There are no specification changes in the 2250, with the exception of the drive capability of the carry-out pin. The low-state drive is 3.2 mA at 0.2 volt, and it is TTL compatible.

2.4.3   The 8260 Seconds/Minutes/Hours BCD Programmable Timer/Counter

The ICL 8260 is manufactured by Intersil. This device, like the 2250 (or 8250) is basically bcd programmable, the main exception being that its maximum count is 59. It is packaged in a 16-pin dual in-line package.

Functional Diagram

A functional block diagram of the 8260 is shown in Fig. 2-21. This diagram is almost identical to that of the 2250 and 8250. A major difference is the deletion of the 80 output (pin 8) on the second counter. In the tens decade, only 10, 20, and 40 are needed for a count of 59. The outputs for the units decade remain the same as for the 2250 and 8250.

Another difference is the timing of the output pin, which goes low at a count of 60. This pin drives the counter of a second 8260 in seconds/minutes counting applications.

Definition of Pin Functions

All pins of the 8260 are identical in general function to those of the 8250. The timing of the carry-out pin is different, however, as is noted next.

Carry Out (Pin 15). This pin is used to signal an overflow of the counter, corresponding to a count of 60. This output goes low at a count of 60 and is used to drive the counter of a succeeding 8260 via the time-base pin. The output is TTL compatible and can thus be used to drive other forms of logic if desired.

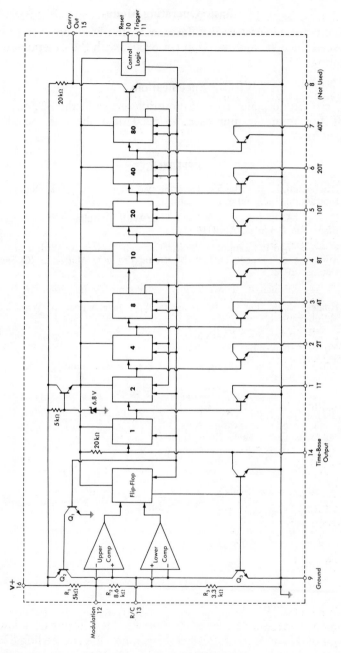

Fig. 2-21. Functional block diagram of the 8260 seconds/minutes/hours bcd programmable timer/counter.

61

Basic Operating Modes

There are no functional differences in the basic operating modes between the 8260 and the 2250 (or 8250), with the exception of the counting.

Specifications

There are no major specification differences in the 8260 over the 8250. For reference purposes, an 8260 data sheet is included in Appendix A.

REFERENCES

1. Camenzind, H. R. "The 555 Timer Story." *Monogram,* No. 2, October 1973. Interdesign, Inc., Sunnyvale, Calif.

2. Grebene, A. B. "Linear IC's Mark Time With a Minimum of Components." *Electronic Products,* April 16, 1973.

3. ———. "The Programmable Timer/Counter—A New IC With a Myriad of Applications." *Proceedings of Electronic Products Magazine IC Seminar,* October 1973.

4. ———. "Which IC Timer to Buy." *Electronic Design,* February 1, 1974.

5. Hnatek, E. R. "Put the IC Timer to Work in a Myriad of Ways." *EDN,* March 5, 1973.

6. Idnani, C. "An Integrated Circuit Dual Timer." *IEEE Transactions,* BTR-20, No. 4, November 1974.

7. Jung, W. G. "The IC Time Machine." *Popular Electronics,* November 1973, January 1974.

8. Mattera, L. "IC Timers Make the Most of Delay." *Electronics,* June 21, 1973.

9. Nelson, C. *Versatile Timer Operates From Microseconds to Hours.* National Semiconductor Application Note AN-97, December 1973. National Semiconductor Corp., Santa Clara, Calif.

10. "Time . . . . IC Controlled." *QST,* June 1972.

11. *Timer/Counter Applications Brochure.* Intersil, Inc., Cupertino, Calif., Fall 1975.

12. Wyland, J. "A Solid State Timer Device for the Consumer." *WESCON Proceedings,* August 1973.

13. ———; Hnatek, E. R. "Unconventional Uses for IC Timers." *Electronic Design,* June 7, 1973.

3

General Operating Procedures and
Precautions in Using IC Timers

In this final chapter of Part I, we are but one step removed from applying IC timers in actual circuits. However, in a practical sense, this chapter is of great importance because it will enable the reader to reap maximum benefit from a given device by optimizing its performance and minimizing possible application pitfalls.

3.1 STANDARD PINOUTS AND TERMINAL DESIGNATIONS

It is a great aid to the understanding of schematic representations of timer circuits if they all follow a standard pattern with common terminology. This section defines the schematic symbol and terminal designations to be used with each device throughout the book. Each timer will appear in the same manner, regardless of the circuit, and the shorthand terms for the various terminals noted here will be used. Wherever possible, like functions of different devices will use similar terminology and symbols for consistency.

3.1.1 555/556 General-Purpose Timers

The 555 and 556 timers, both being general-purpose types, use identical pin designations. They are shown in Fig. 3-1. Fig. 3-1A is the single-unit 555, while Fig. 3-1B is the dual-unit 556 (sections A and B).

In the legend, the pin terminology is shown along with the shorthand form that will be used within the schematic symbol. Thus, the reader will always know that TR is the trigger pin, OUT is the out-

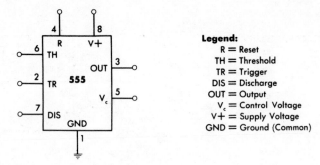

Legend:
- R = Reset
- TH = Threshold
- TR = Trigger
- DIS = Discharge
- OUT = Output
- V_c = Control Voltage
- V+ = Supply Voltage
- GND = Ground (Common)

(A) Type 555 (8-pin package).

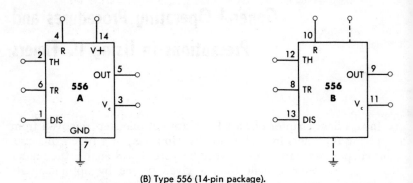

(B) Type 556 (14-pin package).

Fig. 3-1. Pinouts and terminal designations for the 555/556 timers.

put pin, etc. As a further aid to understanding, these pins will always appear in locations similar to that shown.

3.1.2 322/3905 Monostable Timers

The pinouts and terminal designations for the 322 and 3905 monostable timers are as shown in Fig. 3-2. There are several differences in terminology as well as slight differences between the two devices. As shown, the 322 is in a 14-pin dual in-line package and uses 10 of the available 14 pins. Pins used on the 322 that are not common to the 3905 are V_{adj} and boost. The 3905 is in a low-cost, 8-pin package, and consequently does not make these functions available. For all other considerations, however, the two devices are functionally and electrically identical. The schematic symbol is the same for the two devices except for the pin availability and numbering used.

3.1.3 2240/2250/8260 Programmable Timer/Counters

The pinouts and terminal designations for the 2240, 2250, and 8260 programmable timer/counters are as shown in Figs. 3-3, 3-4,

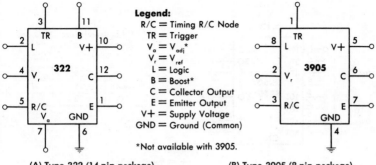

Legend:
R/C = Timing R/C Node
TR = Trigger
$V_a = V_{adj}$*
$V_r = V_{ref}$
L = Logic
B = Boost*
C = Collector Output
E = Emitter Output
V+ = Supply Voltage
GND = Ground (Common)

*Not available with 3905.

(A) Type 322 (14-pin package).   (B) Type 3905 (8-pin package).

Fig. 3-2. Pinouts and terminal designations for the 322/3905 timers.

and 3-5, respectively. All three devices are supplied in a 16-pin dual in-line package.

3.2 TIMING COMPONENT CONSIDERATIONS

With the potential high precision of most of the timing circuits discussed in this book, one of the most important considerations in a design is the selection of appropriate timing components for R_t and C_t. In fact, if this step is not done carefully, the inherent performance capability of the circuits may well be negated. However, when due consideration is given to all factors involved, a design with satisfactory performance can be realized at a reasonable cost.

3.2.1 Resistors

Of the two timing components, resistor selection is by far the easier of the two, for two reasons. First, there is a greater number of

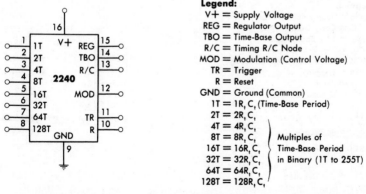

Legend:
V+ = Supply Voltage
REG = Regulator Output
TBO = Time-Base Output
R/C = Timing R/C Node
MOD = Modulation (Control Voltage)
TR = Trigger
R = Reset
GND = Ground (Common)
1T = 1R_tC_t (Time-Base Period)
2T = 2R_tC_t
4T = 4R_tC_t
8T = 8R_tC_t ⎫
16T = 16R_tC_t ⎬ Multiples of
32T = 32R_tC_t ⎪ Time-Base Period
64T = 64R_tC_t ⎪ in Binary (1T to 255T)
128T = 128R_tC_t ⎭

Fig. 3-3. Pinouts and terminal designations for the 2240 binary programmable timer/counter.

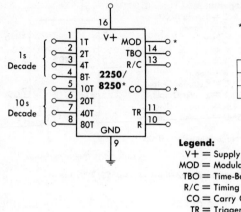

* Pin designations for 2250 and 8250 differ as follows:

	2250	8250
MOD	15	12
CO	12	15

Legend:
V+ = Supply Voltage
MOD = Modulation (Control Voltage)
TBO = Time-Base Output
R/C = Timing R/C Node
CO = Carry Output
TR = Trigger
R = Reset
GND = Ground (Common)

1T ⎫
2T ⎪
4T ⎬ = 1s Decade ⎫
8T ⎭   ⎪ Multiples of
10T ⎫   ⎬ Time-Base Period
20T ⎪   ⎪ in BCD (1T to 99T)
40T ⎬ = 10s Decade ⎭
80T ⎭

Fig. 3-4. Pinouts and terminal designations for the 2250/8250 bcd programmable timer/counter.

standard resistor values from which to choose; this is true whether ordinary 5% types or precision 0.1% types are used. Second, resistor price is not a major function of ohmic value; that is, most values are available at the same price (except at the low-value and high-value

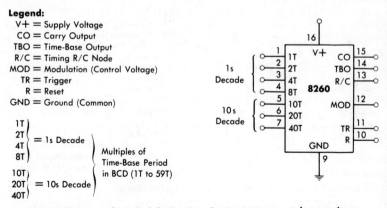

Legend:
V+ = Supply Voltage
CO = Carry Output
TBO = Time-Base Output
R/C = Timing R/C Node
MOD = Modulation (Control Voltage)
TR = Trigger
R = Reset
GND = Ground (Common)

1T ⎫
2T ⎪
4T ⎬ = 1s Decade ⎫
8T ⎭   ⎪ Multiples of
10T ⎫   ⎬ Time-Base Period
20T ⎬ = 10s Decade ⎪ in BCD (1T to 59T)
40T ⎭   ⎭

Fig. 3-5. Pinouts and terminal designations for the 8260 seconds/minutes/hours programmable timer/counter.

extremes). Thus, it is fairly certain that a wide range of values will be available for any given resistor type.

In the class of general-purpose resistors, both carbon-composition and carbon-film types are available. These are generally supplied in tolerances of 5% and 10%, with some carbon-film units in 2% tolerances. Temperature coefficients vary depending on the exact type and value but are in the range of 200 to 500 ppm/°C. These types of resistors are suitable for breadboarding or for general-purpose noncritical applications where a total range of resistance uncertainty and/or instability can be 10% or more.

A higher-quality resistor is the metal-film type available from a large number of manufacturers. These units are supplied in a much broader range of values, even down to and including 1% increments if desired. Initial tolerances can be from 0.1% to 1%, and TCs are available from 25 to 100 ppm/°C. It is this type of resistor that is most suitable for precision or semiprecision timers. Their cost, while higher than that of carbon resistors, is reasonable when performance is considered.

In general, the foregoing statements are true for resistor values from 100 Ω to 1 MΩ. Unfortunately, for values above a few megohms, precise and stable resistor types become harder to procure and are more expensive. When accuracy and stability are required in the range of 1 MΩ to 100 MΩ, the unit selected will most likely not be a stock component.

Applications that require long timing periods should be examined very closely for possible tradeoffs between the resistor and the capacitor. Also, a timer/counter to extend the timing range should be considered. This latter option can often be a workable solution in view of the cost and procurement problems of high-value resistors and capacitors.

Adjustment is often desirable in the timing resistor. If not handled carefully, however, the introduction of a potentiometer can sacrifice the reliability of a timing circuit. From a reliability standpoint, it is good practice to avoid potentiometers if at all possible. Of course, there are situations that demand them, such as panel-operated circuit controls. In these cases, it is always best to select a high-quality control, due to the frequency of adjustment, operational "feel," environment, etc.

Circuit-trimming potentiometers can also be sources of trouble. In general, the percentage of resistance range of a trimmer should be minimized to just in excess of that required. For example, a potentiometer that has a relatively poor TC will have its TC in the circuit improved by a factor of ten if the trim range is reduced to 10%. The open-element carbon-type trimmer should be avoided, if possible, unless performance requirements are modest. The multiturn cermet types provide excellent performance for their cost and are available in a wide range of values.

Regardless of the resistor type used, its stability will be enhanced if it is well derated with regard to power. This will usually not be a problem except for low-value resistors at the higher supply voltages. A good goal is 1/5 or less of rated power, rather than the usual "half-power" reliability rule.

3.2.2 Capacitors

Capacitor selection is by far the biggest problem of all in the design of a timing circuit, for several reasons. These reasons include the necessity for understanding the many capacitor types, their performance/cost tradeoffs, and the severe limitations of capacitors for timing applications, particularly in the larger values. Capacitors are by their very nature a somewhat imprecise component, and few types are even available that have tolerances below 5%. Of those types that are available, not all are suitable for timing applications, for one reason or another. In addition, the range of available values is limited.

In general, for timing circuit use, a capacitor must have very low leakage, good dielectric-absorption characteristics, and a low temperature coefficient for stability. The low-leakage point is an obvious one; if a timer is to use a capacitor that is to be charged from a 1-μA source, it must have a self-leakage much lower than this. Furthermore, the leakage must remain low over conditions of voltage, environment, and time. The second point is particularly important for timing-circuit capacitors. Some dielectrics used in capacitors exhibit the phenomenon known as *dielectric absorption* (abbreviated DA). Very simply, this means that when the capacitor is charged and then discharged by shorting the leads together, the dielectric does not give up all of the energy that was stored when the capacitor was charged. The capacitor is then said to exhibit a *residual* voltage. For timing circuits, this is an obvious drawback because the timing principle depends upon starting from zero. And, in some capacitor types, the DA (measured in percent) can be as high as several percent, so it is a very real problem.

Capacitor types with high dielectric absorption should be avoided for timing circuits if good performance is desired. These include papers, ceramics, and some mica types. All of these types can have DAs of 3% to 5%. Impregnated dielectrics will generally be poor in terms of dielectric absorption, although they may excel in other characteristics. There is one family of capacitor types, however, that is very good in terms of dielectric absorption, and that is the plastic-film capacitors. These include polystyrenes, polycarbonates, and some others.

Polycarbonate types are below 1% for dielectric absorption, and polystyrene and parylene types exhibit a DA below 0.1%. Teflon* is another excellent capacitor dielectric for timing circuits, its only drawback being its relatively high cost.

Polystyrene is one of the best dielectrics in terms of cost and performance, but it does have two limitations. It cannot be used above 85°C, and it is available in capacitance values only up to about 1 μF, and this range is not always available from all sources. Advantages of polystyrene include tolerances down to 1% and a very linear TC of −120 ±50 ppm/°C. While not zero, this TC is low enough for many applications; for other applications, the fact that the TC is linear allows compensation with a thermistor if necessary. This TC is the lowest available in low-dielectric-absorption types. Polystyrene capacitors are available from many sources.

Parylene, a proprietary dielectric of Union Carbide, is used in the Kemet F310 series of plastic-film capacitors. These types range in value from 0.001 μF to 1 μF with tolerances down to 0.5%. Their TC is linear at −200 ±30 ppm/°C. This capacitor type is useful over a temperature range of −55°C to +125°C.

Polycarbonate is the next most suitable dielectric and is widely available from a number of suppliers at moderate cost. The values range up to several tens of microfarads from some suppliers, and close-tolerance units are available. Temperature coefficient in this type is not as linear as in the polystyrene or parylene types; thus it cannot be compensated as easily. In general, types can be obtained that will exhibit a total value change of less than 1% over a 0°C to 70°C temperature range, without compensation.

Although this discussion obviously involves low-voltage capacitor types (i.e., 25 to 50 Vdcw), we are nevertheless still talking about a relatively large physical device for capacitances of 1 μF or above (i.e., 1 to 2 inches in length). This is, unfortunately, one of the capacitor "facts of life." To some extent, capacitor size in a given application can be minimized by increasing the value of the timing resistor; in fact, this tradeoff should always be made because capacitor price is a strong function of size (and not necessarily a linear one). Where possible, timer/counters can help minimize capacitor size, as previously mentioned.

One type of capacitor that has not as yet been mentioned is the electrolytic type. This is simply because electrolytics are far from optimum for timing-capacitor use. Their drawbacks are: very loose tolerances, poor stability, and high leakages. In general, aluminum electrolytics are not suitable at all for timing circuits, unless the application is one of low-performance requirements.

---

* A registered trade name of E. I. duPont de Nemours & Co., Inc.

Tantalum electrolytics can be used with certain qualifications. For limited temperature ranges such as 0°C to 50°C, tantalums are very attractive economically and offer reasonable performance. One example of this type is the Union Carbide Kemet T310 series. In general, voltage derating will aid in controlling leakage in these types. Since leakage can be as high as several microamperes in tantalum types, a reduction of usable resistance range will accompany their use. Timing current should be several times greater than the highest anticipated capacitor leakage.

For reference, a list of suppliers of suitable timing resistors and capacitors is included in Appendix C.

3.3 GENERAL DESIGN PRECAUTIONS

In applying IC timers there are a few general precautions that should be observed; these may vary slightly from device to device. These considerations fall under the category of good design practices.

The power supply may need some attention depending on the particular timer being used. Although 555-type devices are by design relatively immune to changes in timing caused by supply-voltage variations, they are sensitive to power-supply changes that occur *during the timing period*. This factor may be dealt with by appropriate bypassing, either at the control pin or at the supply terminal. In stubborn cases, or in the presence of severe supply-line noise, decoupling may be necessary in the form of a local zener regulator of 5 to 10 volts.

Even if the supply line is relatively clean, a 555-type timer can in itself generate supply noise. This is because of the totem-pole output stage, which causes current spikes in the power line when it changes states. For this reason, it is usually good practice to provide some bypassing at the IC or at least on the printed-circuit card. Low-inductance capacitors such as $0.01\text{-}\mu\text{F}$ to $0.1\text{-}\mu\text{F}$ ceramics or $1\text{-}\mu\text{F}$ solid tantalums are recommended for this.

In general, attention should be given to potential noise sources that may influence timing accuracy in or around the comparator inputs, particularly when high-impedance timing networks are being used. Timing-capacitor returns should be connected directly to the timer ground pin to avoid common ground currents, which could influence the timing voltage.

When high-impedance timing networks are used with low-input-current devices such as the 322 or 3905, attention should be given to leakage paths to the timing pin. Avoid possible stray leakage coupling to high potentials. In addition, the circuit board should be cleaned with an appropriate chemical cleaner, and then sealed against humidity.

All of the timers discussed in this book are single-supply devices and are normally used as such. There are instances, however, when one or more of the timer pins are interfaced with external active circuitry such as an op amp. In these cases, overvoltage or reserve voltage modes (even on a transient basis) should be carefully evaluated and positively prevented. This can take the form of simple current limiting or diode clamping and is not necessarily complex. In general, the "safe limit" rule of each timer pin must be observed for reliability.

3.4 PERIPHERAL DEVICES

This section briefly discusses several IC components that are generally useful as peripheral devices operating in conjunction with IC timers. All of the devices discussed here and in the applications section have one or more common features of compatibility with timers. Among these are similar supply-voltage ranges, single-supply operation, simple and predictable operation, and/or high performance at low cost.

3.4.1 Op Amps

One general class of devices useful for augmenting timer operation is that of the operational amplifier (op amp). Op amps are useful with timers for buffering the inputs and/or creating various related control functions. Many different op-amp types are usable with timers, but a few stand out as being particularly useful, either because of technical features or because of a unique combination of cost and performance. These devices are illustrated schematically in Fig. 3-6.

The most popular and widely known op amp is the 741, which is shown in Fig. 3-6A. A 741 may be operated from either dual or single power supplies, but it is not as useful as some others with single-supply operation below 10 volts. It can be nulled for greater accuracy via R_{adj}, as shown in dotted lines. Offset nulling is at the option of the user and can be deleted if simplicity is required. This is generally true of many other devices, and their recommended optional null networks are also shown in dotted lines.

One device that is particularly suitable for use with timers is the 3140, an op amp with a FET input stage, which allows an extremely low input current of only 10 pA. The 3140 is suitable for single-supply operation from below +5 volts to +36 volts, or it can be used with dual supplies. The supply-voltage ranges listed are those of the basically recommended device—in this case, the 3140. The uppermost voltages (shown in parentheses) are for single-supply operation, while the lower voltages are for dual power supplies.

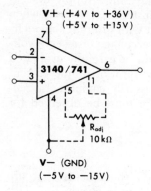

(A) Types 3140/741.

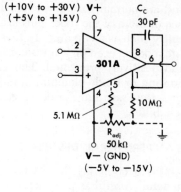

(B) Type 301A.

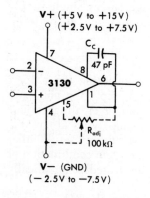

(C) Type 3130.

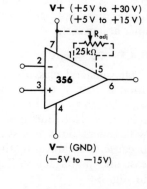

(D) Type 356.

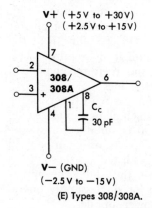

(E) Types 308/308A.

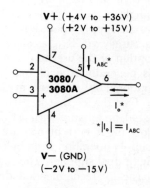

(F) Types 3080/3080A.

Fig. 3-6. Timer-compatible op amps.

The 3140 also has the ability to operate with the input at zero volts (ground) when operated from a single supply. This one feature is highly desirable for use with timers, as it eliminates the need for the dual supplies usually required with op amps. The combination of features mentioned makes the 3140 probably the single most useful op amp for timer use. The pinouts for the 3140 are identical to the 741, as seen in Fig. 3-6A. This particular pin configuration (with the nulling) is the most standard one among op amps and is applicable to many others in addition to the 741 and 3140. Some examples are the 8007, 1456, 1436, and 776, to name just a few.

Another popular general-purpose op amp is the 301A, which is shown in Fig. 3-6B. Pin configuration is similar to the 741, but an additional compensation capacitor (C_C) is necessary. This hookup is also suitable for the 748 and 777 types.

The 3130 (shown in Fig. 3-6C) is another FET-input op amp, one which preceded the 3140. It is limited to a total supply voltage of 15 volts, but also has the ability to operate with the input at zero volts when operated from a single supply. It requires a compensation capacitor (C_C) and has the unique characteristic of an output stage that can swing to each supply limit. It can be used from +5 volts to +15 volts; thus it is directly compatible with 555-type timers.

The 356 (shown in Fig. 3-6D) is also a FET-input amplifier but with dc performance specifications that place it in a more premium class than the 3130 or 3140. It is suitable for higher-accuracy applications that demand high stability in terms of input specifications. It is nulled differently, as shown.

Pinouts for the 308-type op amps are shown in Fig. 3-6E. As may be noted, this amplifier is compensated identically to the 301A. The main difference is that there is no provision for offset nulling in the 308 types. This is not necessarily a disadvantage as the 308A features a maximum offset voltage of 500 μV, plus other premium specifications such as a typical bias current of 1.5 nA. These devices can operate from single supplies as low as 5 volts and possess the highest dc accuracy of the types listed here.

Pinouts for the 3080 amplifier types are shown in Fig. 3-6F. This device is not a conventional op amp but is an operational transconductance amplifier (OTA). The output is in the form of a current, I_o, the maximum value of which is programmable by means of current I_{ABC} into a third input terminal. The direction of output current flow (into or out of the output) is determined by the differential inputs. The 3080A is useful over a programming current range of more than five decades, and both types are useful over wide supply-voltage ranges.

Quad and dual op-amp types are also available and are useful in systems applications. Pinouts for the 324 and 3403 quad op amps are

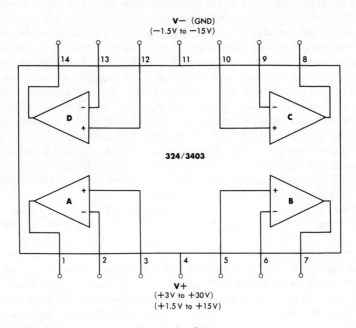

(A) Types 324/3403.

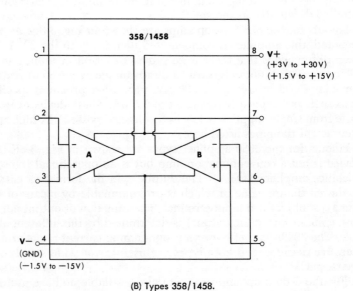

(B) Types 358/1458.

Fig. 3-7. Quad and dual op-amp types.

shown in Fig. 3-7A. Both these devices operate from single supplies from below +5 volts to +30 volts, and they will operate with the inputs (and outputs) at zero volts. Their input bias current is much higher than the 3130, 3140, or 356 types, however. They are internally compensated and have no offset-null provision. Of the two units, the 324 is more useful to timer applications for several reasons. It has a very low (standby) supply current (0.8 mA), and its input bias current is much lower than the 3403 (45 nA). Both devices have a wide supply-voltage range.

The 358 op amp, shown in Fig. 3-7B is a dual device in an 8-pin package. Its electrical performance is identical to the 324. The 1458, which has identical pinouts, is simply two 741-type devices. The supply-voltage range shown is for the 358 only; the 1458 is subject to the same supply-voltage and input/output restrictions as a 741.

For further insight into these devices, and op amps in general, references are listed at the end of the chapter.

3.4.2 Logic Devices

Logic devices that are compatible with timers could include virtually all of those available, which number in the thousands. However, certain devices interface more readily with timers than others, due to supply-voltage range, threshold levels, and/or loading considerations. For instance, if the timer circuit is operated from a 5-volt supply, the entire 54/74 series of TTL logic is available to the designer for interfacing, at both the timer input and output. A discussion of the hundreds of TTL devices available is beyond the scope of this book, however, and the reader is referred to the references at the end of the chapter.

One logic family that is highly compatible with timers is the CMOS family, as it has an identical supply-voltage range of +5 volts to +15 volts. Furthermore, CMOS inputs are essentially nonloading, as their input current is only 10 pA. They have a logic threshold of ½ of V+; therefore, they may be driven from virtually any timer output without concern for interfacing problems. Also, the very low leakage characteristic of CMOS switching elements makes them ideal for direct use within the RC timing section, without compromise in timer performance. Finally, the speed capability of CMOS, while far from the highest by modern logic standards, is more than adequate in terms of compatibility with timers. At low speeds and/or long timing periods, the virtually zero power dissipation of CMOS is a strong point in its favor. It is for these reasons that CMOS devices are used within the applications section.

A few CMOS devices that are compatible with timers are shown in Fig. 3-8. These devices are from the CD4000 Series, a logic family introduced by RCA and second-sourced by many others.

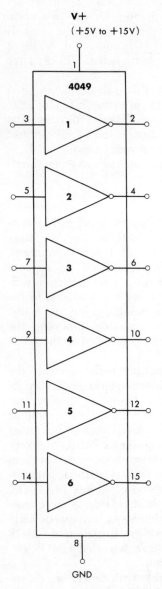

(A) Type 4049 hex inverter.

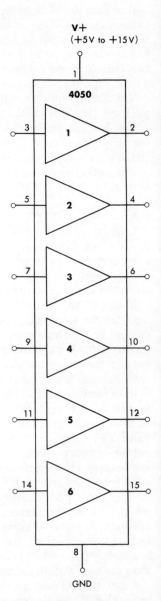

(B) Type 4050 hex buffer.

Fig. 3-8. Timer-compatible

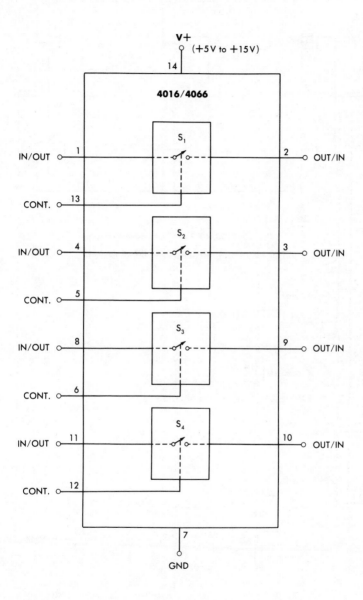

(C) Types 4016/4066 quad bilateral switches.

CMOS devices.

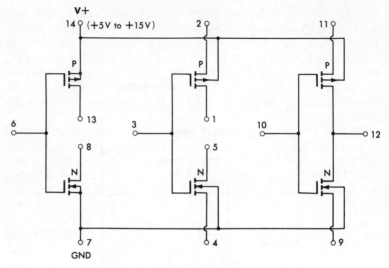

(D) Types 3600/4007 CMOS transistor array.

Fig. 3-8 (cont). Timer-compatible CMOS devices.

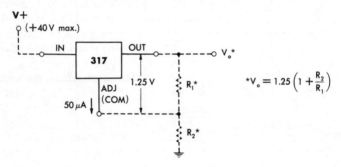

(A) The 317 as basic voltage regulator.

$$*V_o = 1.25 \left(1 + \frac{R_2}{R_1}\right)$$

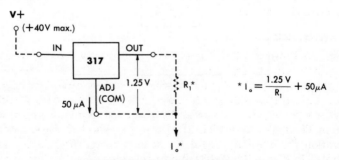

(B) The 317 as basic current regulator.

$$*I_o = \frac{1.25\,V}{R_1} + 50\mu A$$

Fig. 3-9. Type 317 voltage/current regulator.

For buffering and logical inversion of timing signals, the 4049 hex inverter shown in Fig. 3-8A is very useful. This device can be operated from +5 volts to +15 volts as a general-purpose inverter but may also be used as a logic level converter, with a 5-volt supply and a 15-volt input as one example. This device is very useful in such cases as driving a TTL system from a 15-volt timer output. The 4050, shown in Fig. 3-8B, is a device that operates similarly but does not invert the signal.

The 4016, shown in Fig. 3-8C, is not a logic element in the strictest sense but is actually a set of four logically controlled switches. Switches S_1–S_4 are CMOS transistors whose conduction state is controlled by input control lines. Switch S_1, for example, is "on" when pin 13 is high and is "off" when pin 13 is low. The switches are bilateral and may be referred to any voltage between zero and the V+ level used. The "on" resistance is typically 300 Ω at a V+ of 15 volts, and off-state leakage is typically 100 pA. A device that has identical pinouts but features a lower "on" resistance is the 4066. The R_{on} in this device is typically 80 Ω at a V+ of 15 volts; otherwise it is similar to the 4016.

Another device that can be quite useful in timing circuitry is the 4007, shown in Fig. 3-8D. This unit is an array of uncommitted n- and p-channel MOS transistors, which can be wired for a variety of functions. Gate drive to any input is the typically low CMOS current of 10 pA. This same basic circuit structure is also available in the RCA CA3600, which is characterized for linear service.

There are, of course, many more CMOS devices that are suitable for use with timers. One notable family is the 54C/74C series by National Semiconductor, which is also second-sourced by others. This CMOS family is functionally equivalent to, and pin-compatible with, the 54/74 TTL types, but with CMOS characteristics. Functions include most of the TTL series counterparts, and some not found in the CD4000 series.

For further background on CMOS devices, references are listed at the end of the chapter.

3.4.3 Regulators

Fig. 3-9 illustrates the 317, a 3-terminal, positive voltage-regulator IC, which is useful with or within timer circuits. This device is a programmable output regulator with a current capability of up to 1.5 A and an output voltage range of +1.25 volts to +37 volts. It may be used as either a voltage or a current regulator and is available in TO-3, TO-5, and TO-220 packages. The two basic modes of operation (as a voltage regulator or as a current regulator) are illustrated in Figs. 3-9A and 3-9B, respectively. Salient points of device operation are the 1.25-volt ±4% terminal voltage between the OUT

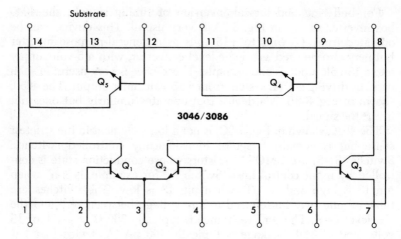

Fig. 3-10. The 3046/3086 IC transistor array.

and ADJ terminals, and the 50-μA bias current flowing from the ADJ terminal.

3.4.4 Transistor Arrays

A very convenient IC for use with timer circuits is the monolithic transistor array. This type of device, which is available in a variety of configurations, is a great aid to circuit design from the standpoint of matching device characteristics and packing density. A popular representative type is the 3046 array shown in Fig. 3-10. These five transistors have general-purpose specifications, but they also feature matching characteristics and thermal tracking. The 3086 is an array that has identical pinouts but with slightly different specifications.

REFERENCES

1. Calebotta, S. *CMOS, The Ideal Logic Family.* National Semiconductor Application Note AN-77. National Semiconductor Corp., Santa Clara, Calif.

2. Dansky, S.; Funk, R. E. *Handling and Operating Considerations for MOS Integrated Circuits.* RCA Application Note ICAN-6000, March 1974. RCA Solid State Div, Somerville, N.J.

3. Havasy, A.; Kutzin, M. *Interfacing COS/MOS With Other Logic Families.* RCA Application Note ICAN-6602, November 1973. RCA Solid State Div., Somerville, N.J.

4. Johnson, F. L. *Capacitors . . . Dielectric Absorption.* Technical Bulletin No. 10. Electrocube Inc., San Gabriel, Calif.

5. Jung, W. G. "A Guide to CMOS Operation." Part 1, *Popular Electronics*, March 1974. Part 2, *Popular Electronics*, April 1974. "How CMOS Devices Are Used in Linear Applications," *Popular Electronics*, August 1974.

6. ———. *IC Op-Amp Cookbook*. Howard W. Sams & Co., Inc., Indianapolis, 1974.

7. Lancaster, D. *TTL Cookbook*. Howard W. Sams & Co., Inc., Indianapolis, 1974.

8. Melen, R.; Garland, H. *Understanding CMOS Integrated Circuits*. Howard W. Sams & Co., Inc., Indianapolis, 1975.

9. Redfern, T. P. *54C/74C Family Characteristics*. National Semiconductor Application Note AN-90. National Semiconductor Corp., Santa Clara, Calif.

10. Manufacturer's Data Sheets:

Motorola	{	MC3403	Operational Amplifier.

National Semiconductor	LM301A	Operational Amplifier.
	LM308	Operational Amplifier.
	LM308A	Operational Amplifier.
	LM324	Operational Amplifier.
	LM356	Operational Amplifier.
	LM358	Operational Amplifier.
	LM317	Voltage/Current Regulator.
	MM54C/74C	Series, Logic Devices.

RCA	CA3600	CMOS Transistor Array.
	CA3046	NPN Transistor Array.
	CA3086	NPN Transistor Array.
	CA3130	Operational Amplifier.
	CA3140	Operational Amplifier.
	CA3080	Operational Transconductance Amplifier (OTA).
	CD4000	Series, Logic Devices.

11. Military Specifications:

MIL-C-19978D—Capacitor, Fixed, Plastic (or Paper-Plastic) Dielectric (Hermetically Sealed in Metal, Ceramic, or Glass Cases), General Specification For.

MIL-C-55514A—Capacitor, Fixed, Plastic (or Metallized Plastic) Dielectric, DC, in Nonmetal Cases, General Specification For.

II

IC TIMER APPLICATIONS

4

Monostable Timer Circuits

This chapter is the first to deal with practical timer circuitry of other than a general nature. Specifically, it is concerned with the operation of various types of monostable timing circuits, some of which are but slight modifications of the basic monostable circuits discussed earlier; others show extensive modifications. All of the circuits, however, serve some specific purpose. Not only are they useful on a "stand-alone" basis, but many of them can also be modified further to form parts of larger-scale systems, which is the subject of Chapter 6. Pertinent design information is given for the circuits, with illustrations. Also, the limits of performance are discussed, particularly when different from the standard form.

In general, the devices utilized are those most suitable for monostable operation. Comments pertinent to a 555 circuit are equally applicable to one-half of a 556, with proper attention given to pin numbering, of course. Similarly, many comments pertinent to a 322 can also apply to a 3905 (or vice-versa).

In the interest of clarity, some simplifications have been made in the treatment of schematics. Supply voltage conditioning and by-passing are not specifically shown with each circuit, but it should be understood that they are implicitly necessary (at least to some degree) with all of them. The reader is referred to Chapter 3 for specific guidance on each device in this regard.

Many of the circuits operate without qualifications over the full supply-voltage range of the IC device(s) listed. When this is the case, a general supply-voltage notation (simply V+) is shown. This should be understood to mean the full supply-voltage range for the specific device listed, such as the 555, for example, which operates from +5 volts to +15 volts. Where a circuit operates optimally at some specific voltage (or range of voltages), much as +15 volts rather than the general V+, this is listed. In such cases, the limitations of the circuit will be discussed further within the accompanying text.

Also in the interest of overall clarity, the input trigger conditioning circuitry is not shown. The reader is cautioned, however, to provide satisfactory trigger signals for the device being used (see Chapter 2 for details of device triggering requirements).

4.1 555 MONOSTABLE WITH AUXILIARY OUTPUT

A circuit that is only slightly different from the basic 555 monostable is shown in Fig. 4-1. This circuit has the same input trigger requirements and timing equation but differs in that two outputs are available rather than one, as in the conventional case. The difference lies in the removal of pin 7 as a clamp for C_t in the standby state, and also as a discharge source for C_t. This frees pin 7 for external circuit use as an uncommitted open-collector output. The advantage of using pin 7 in this manner is that it may be referred to any supply voltage between zero and +15 volts, regardless of the timer's supply voltage. As a consequence, flexibility is enhanced.

During the quiescent state of this circuit, the output will be low and C_t will be discharged by virtue of being effectively in parallel with R_t. When the circuit is triggered, the output goes high, and C_t is charged through R_t. The timing cycle is conventional, and the pulse width is $T = 1.1 R_t C_t$, as in the standard case. When the

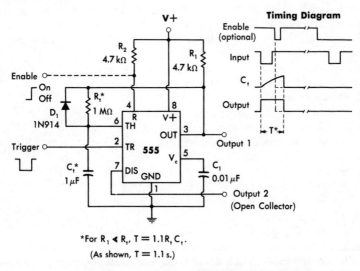

*For $R_1 \blacktriangleleft R_t$, $T = 1.1 R_t C_t$.
(As shown, $T = 1.1$ s.)

Fig. 4-1. A 555 monostable circuit with auxiliary output.

voltage across C_t reaches the threshold of pin 6, the output goes low, ending the timing cycle. This discharges C_t through D_1 initially, until the voltage decreases to the contact potential of D_1 (0.6 to 0.7 volt). Capacitor C_t then discharges more slowly through R_t to its final level between pulses.

There are some weaknesses to the circuit, primarily involving the different manner of charging and discharging C_t. An output pull-up resistor (R_1) is added to force the output completely up to the V+ level in the high state. The value of R_1 should be lower than R_t for minimum error in timing, if it is not actually included as part of R_t in the timing cycle calculations.

Diode D_1 can effectively and rapidly discharge C_t, but only down to 0.6 to 0.7 volt, when it (D_1) ceases to conduct. After D_1 turns off, C_t must then discharge through R_t. The period of time required to discharge C_t completely is longer than in the conventional monostable circuit, and is the major drawback of this circuit. This can cause timing error and/or pulse-width change if the circuit is retriggered just after a completed output pulse. The circuit is thus not optimum for high-duty-cycle operation.

Like the basic 555 monostable, this circuit can be enabled by a high voltage level (or pulse) to pin 4 or inhibited by a low voltage level. Also, a low voltage level (or pulse) to this pin during the timing cycle will terminate or reset the output prior to the normal conclusion of the timing cycle. These characteristics of reset operation are identical to those of the standard 555 monostable, and the use of resetting is entirely optional.

4.2 INVERTED MONOSTABLE*

Fig. 4-2 illustrates a circuit that features an entirely "inverted" sense of operation from the basic 555 monostable. Where the standard monostable accepts negative triggers and delivers positive output pulses, this circuit accepts positive triggers and delivers negative output pulses. As a bonus, it also has an auxiliary output (pin 7) for discretionary use. The key to operation is the fact that the two comparators of the 555 are referenced to 1/3V+ and 2/3V+. When the functions of the trigger and threshold pins are reversed and the timing is referenced to V+ rather than ground, the symmetry of the circuit allows similar pulse-forming operation and an identical timing equation.

In the quiescent state of this circuit, the trigger input (pin 6) is held low (externally) and the output rests at V+. Capacitor C_t is

---

* W. G. Jung, "555 One-Shot Circuit Features Negative Output With Positive Triggering," *Electronic Design*, August 16, 1976.

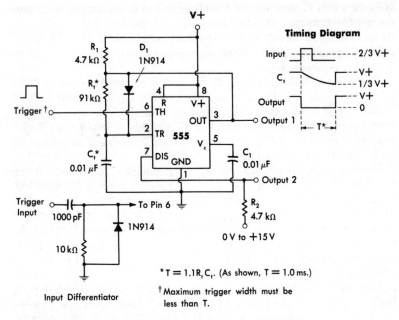

Fig. 4-2. An "inverted" 555 monostable circuit.

charged to V+ through D_1, R_1, and R_t. When a trigger pulse of a level greater than 2/3V+ (the pin 6 threshold) arrives, the output switches low and the timing cycle begins. C_t now discharges toward ground through R_t, as D_1 is reverse biased. Discharge continues until the C_t voltage reaches 1/3V+, the pin 2 threshold. At this point the output switches high again, ending the pulse. Capacitor C_t is then recharged to V+ through D_1, R_1, and R_t.

Timing of the output pulse is identical to the standard 555, and the pulse width is $T = 1.1R_tC_t$. This is because the charging voltage and threshold relationships are maintained the same, even though the sense is inverted. Trigger input requirements are similar to the basic 555 circuit, but inverted. The trigger must be held below 2/3V+ for standby and raised above 2/3V+ for triggering. The trigger must not be held higher than 2/3V+ for a period of time longer than the pulse width. Practically speaking, if the input pulse to be used is wider than T, it should be differentiated (just as is true for the basic monostable). A suitable input differentiator circuit is shown in Fig. 4-2.

One weak point of the circuit lies in the recharging of C_t, which is accomplished initially by D_1 but is charged to final value by R_1 and R_t. This causes the voltage on C_t to slowly approach final value, which will alter the timing if sufficient time is not allowed for recovery. Thus, this circuit will exhibit some timing change or jitter at high duty cycles. An alternate form of this circuit, which eliminates this problem, is described next.

4.3 IMPROVED INVERTED MONOSTABLE

Since the inverted 555 monostable is such a useful design in general, a more accurate and predictable version is justified. This is shown in Fig. 4-3. The operation of this circuit is identical to the basic form, as is the timing expression. The only difference is that an improvement has been made in the charging network for C_t.

In this circuit, the diode has been removed and Q_1, an n-channel JFET switch, has been added (Q_1 may be either of the two types listed). During standby, this switch is held "on" by the high output state, and the typical "on" resistance of 100 Ω (or less) ensures that C_t is charged to the full V+ level. When the output of the timer is low during the timing cycle, this switch is "off," leaving only R_t to discharge C_t to ground. When the timing cycle ends, Q_1 can rapidly recharge C_t to the full V+ level.

Because of these improvements, timing accuracy is improved and duty cycles of 99% or more are possible without jitter. Operation in this regard is at least equivalent to the standard 555 monostable. Care should be taken in loading at the pin 3 output, as alteration of the low- or high-state voltage can affect timing accuracy. If necessary, heavy loads can be driven from pin 7 without loss of accuracy, or this output can be used exclusively.

4.4 MANUALLY TRIGGERED MONOSTABLES

One very common electronic system requirement is the "bug-free" interfacing of switch contacts to circuitry. However, because of the

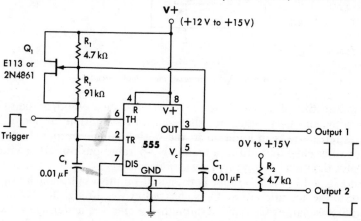

Fig. 4-3. Improved "inverted" 555 monostable.

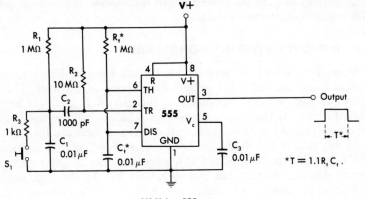

(A) Using 555 type.

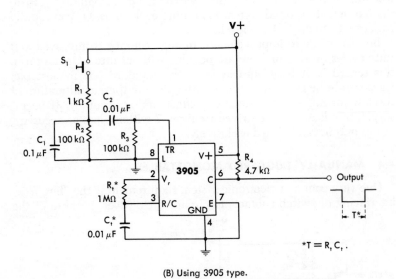

(B) Using 3905 type.

Fig. 4-4. Manually triggered monostable circuits.

characteristics of typical switches (switch bounce and/or noise), this often presents difficulties that can create multiple or sporadic undesired outputs in addition to the desired single output. In Fig. 4-4, two circuits that utilize a manually operated push-button switch to solve the interface problem are shown.

In Fig. 4-4A, the 555 timer is triggered by push-button switch S_1. Prior to the actuation of S_1, C_1 is charged to V+ through R_1. Depressing S_1 discharges C_1 rapidly through R_3, creating a short negative spike. Any sporadic effects of switch bounce that may occur when S_1

is depressed are removed by the integrating action of R_1 and C_1. The resultant clean negative spike is then passed through C_2 to the 555 as a trigger pulse. This fires the monostable, generating an output pulse of width $T = 1.1R_tC_t$. Upon release of S_1, C_1 recharges to V+, and the circuit awaits the next switch actuation.

The time constants and circuit arrangement of the trigger circuitry are chosen for a single trigger and output pulse for each switch depression only. Furthermore, the circuit will not trigger on switch release and will produce only one pulse regardless of how long S_1 is held down. The circuit will also retrigger as fast as S_1 can be reactuated.

In some cases, the 555 can be triggered more directly by a simple switch contact from pin 2 to ground. However, this is usually undesirable because the output will remain high after one timing period if the switch is held closed, which is an undefined mode of operation. The input triggering network consisting of R_1, R_2, R_3, C_1, and C_2 eliminates this problem. Although five components are used, none of them are critical.

The minimum output pulse width of this type of circuit should be longer than the bounce of the switch used; typically this will be 10 ms or less. Thus, the circuits shown are 10 ms or more in pulse width.

A circuit arranged for the 3905 timer is shown in Fig. 4-4B. Here a similar trigger network is included, with slightly different values because of the lower trigger-input impedance of the 3905. In operation, R_1 charges C_1 upon closure of S_1, which fires the timer, creating an output pulse of $T = R_tC_t$. Although a negative-going output pulse is shown, it should be recalled that this device can be arranged for either negative or positive outputs, according to the state of the logic input. As in the 555 circuit, the switch and trigger circuit is designed for a single output pulse upon actuation of S_1, and it will not trigger on switch release.

4.5 TOUCH SWITCH*

An interesting type of circuit that uses an IC timer to advantage is shown in Fig. 4-5. This circuit is a "touch switch," a device in which switching action is accomplished without conventional mechanical switch contacts. The circuit is basically a conventional 555 monostable, with the only major difference being its method of triggering. In this case, the trigger input is biased by a high-value (22-MΩ) resistor, R_1. This maximizes the sensitivity of the circuit

---

* J. C. Heater, "Monolithic Timer Makes Convenient Touch Switch," *EDN*, December 1, 1972.

because the input impedance will be essentially that of R_1, and the trigger input will not draw significant current when held above its $1/3V+$ threshold voltage.

With threshold control R_2 adjusted so that the voltage at pin 2 is held above $1/3V+$, the circuit will remain in its quiescent state (output low) prior to triggering. When the contact plate is touched, the body capacitance of the person touching the plate, being effectively in parallel with the high resistance of R_1, will lower the overall impedance between pin 2 and ground sufficiently to reduce the voltage at pin 2 to below the $1/3V+$ trigger threshold. Thus, the timer will be triggered and will time out, producing an output pulse of width $T = 1.1R_tC_t$. The timing period should be made longer than the anticipated contact time, otherwise the timer will retrigger after completion of the first output pulse. In the example shown in Fig. 4-5, the timing period is about 5 seconds.

The contact plate can be any conducting material arranged for convenient finger contact. Also, some type of feedback to the operator is desirable, such as a lamp or LED to indicate that switching has occurred as a result of contact.

4.6 POWER-UP ONE-SHOTS

The need often arises, particularly in digital systems, for a monostable timer that will produce a single, well-defined pulse upon the application of power. This "power-up" pulse is used to initialize various circuits to a desired predetermined state. Both the 555- and

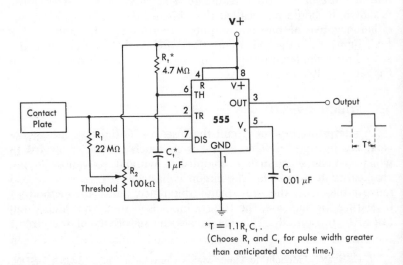

$$*T = 1.1R_tC_t.$$
(Choose R_t and C_t for pulse width greater than anticipated contact time.)

Fig. 4-5. Touch-switch circuit.

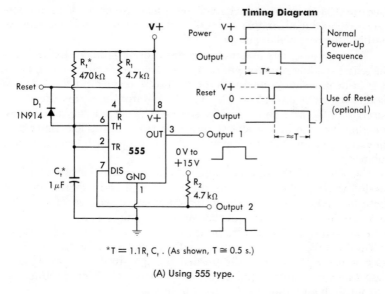

*T = 1.1R_t C_t . (As shown, T ≅ 0.5 s.)

(A) Using 555 type.

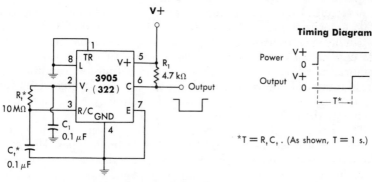

*T = R_t C_t . (As shown, T = 1 s.)

(B) Using 3905 (322) type.

Fig. 4-6. Power-up one-shot circuits.

the 322/3905-type timers can be used efficiently for this function, as shown in Fig. 4-6. In Fig. 4-6A, a 555 circuit that can be used both as a power-up one-shot and as a reset or recycle one-shot is shown. Its use as a power-up one-shot will be described first.

If the threshold and trigger inputs of a 555 are connected together and also to the junction of a timing network, R_tC_t, the device will self-trigger and complete a single timing cycle. This is because at the time of turn-on, C_t is discharged, which momentarily holds pin 2 low, beginning a timing cycle. Capacitor C_t will then charge up to V+ through R_t, and when the voltage crosses 2/3V+, the pin 6

93

threshold is reached, which completes the cycle. A single output pulse is produced in this fashion, and the circuit will not recycle without the removal of power.

The preceding has assumed the reset input (pin 4) to be high, which allows the sequence to occur as described. Once cycled, however, this one-shot configuration can also be recycled, if desired, by taking the reset input low. This forces an output-low condition, and discharges C_t through D_1 and the source. The output will remain low for the duration of the low state at the reset input. The timer will then produce an output pulse when the reset input is taken high once again.

The width of the output pulse produced when the circuit is recycled through the use of the reset input is slightly shorter in time because of D_1, which prevents the complete discharge of C_t. This is usually of small consequence, since the pulses produced by this type of circuit rarely require precision. The use of the reset function is entirely optional, and if not needed, D_1 and R_1 can be omitted. The circuit has two outputs, the normal pin 3 output (output 1) and also the uncommitted pin 7 output (output 2). Both produce an output-high condition during the timing cycle.

A circuit that performs the power-up one-shot function with a 3905 timer is shown in Fig. 4-6B. Both the 322 and 3905 have the capability of producing an output timing cycle automatically upon the application of power. As a consequence, there is no triggering required of the circuit, and it produces an output-low condition for the duration of the timing period after the application of power. Upon completion of the timing cycle, the output goes high and stays high until the power is recycled.

This circuit cannot be recycled with the power on, due to the absence of a reset function on the 322/3905 timer types. However, the circuit does have the ability to produce either high or low outputs during the timing cycle by use of the logic input (pin 2 on the 322; pin 8 on the 3905). The circuit as shown uses the 3905 since the timing interval is much greater than 1 ms, but the 322 can also be used if desired.

4.7 RESTARTABLE ONE-SHOT

During normal operation, a 555 monostable circuit will ignore input triggers that may occur during the timing cycle and will produce uninterrupted output pulses. On occasion, the need may arise to by-pass this mode of operation and to restart the output timing with the application of a trigger pulse during the normal timing period. A circuit that accomplishes this is shown in Fig. 4-7. In this circuit, the reset and trigger pins of the 555 are tied together and used as a single trigger input. When the input is driven to zero, it has reached the threshold level of the reset input. Since the reset function is over-riding, the output is held low even though the trigger input is below its 1/3V+ threshold. When the input rises, the reset function loses control and the trigger input assumes command. The output then rises, and a timing cycle begins.

If no additional trigger pulses are received prior to completion of the timing cycle, an output pulse of normal width will be produced. (See the timing diagram of Fig. 4-7 for normal operation.) The only difference between operation in this circuit and a conventional 555 monostable is that timing begins on the rising (trailing) edge of the input pulse rather than on the falling (leading) edge.

When input triggers are spaced at time intervals less than the normal timing cycle, the cycle will not reach completion and the output will go low with the falling edge of each input pulse. This is shown in the timing diagram for restartable operation. It may be noted that C_t is never allowed to reach the normal voltage level; thus, the output pulses are not controlled by R_t and C_t in this circuit. The output is actually an amplified replica of the input.

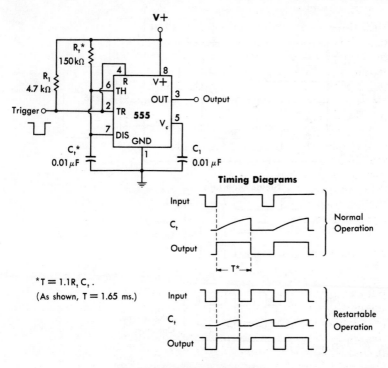

Fig. 4-7. Restartable one-shot circuit.

Appropriate selection of R_t and C_t can control the "crossover" time period when the circuit ceases to function as a normal monostable and begins restartable operation. This may be anywhere within the range of component values allowable for the 555.

4.8 RETRIGGERABLE ONE-SHOT

The circuit of Fig. 4-8 is similar to the circuit of Fig. 4-7 in that it modifies the normal triggering characteristic of a 555-type monostable during an output pulse. There is a difference, however, in the fact that the output of the monostable shown in Fig. 4-8 is not reset, although a new timing cycle is begun. This type of operation is termed *retriggerable*.

In this circuit, which is made up of two halves of a 556, section B is a conventional 555-type monostable with timing components R_{t_2} and C_{t_2}. This section is triggered by the negative-going pulse from section A, which is a simplified inverted monostable. The free output terminal of section A (pin 1) is used to clamp C_{t_2} of section B

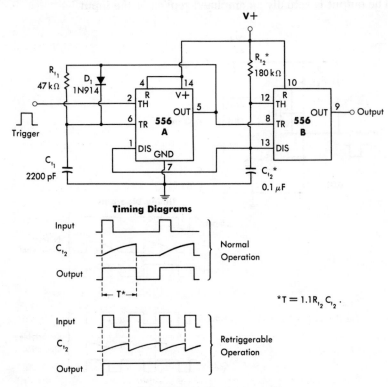

Fig. 4-8. Retriggerable one-shot circuit.

for the duration of the output pulse of section A, which in this case is 100 μs. Input triggers to section A are positive, and initiate both timer sections, since the first section fires the second section.

During normal operation, where the input triggers are received at a time interval longer than the period of monostable B, the output is conventional and is shown in the timing diagram for normal operation. Output pulses occur at the input rate, and the output pulse width is equal to $1.1R_{t_2}C_{t_2}$. When the input period is shorter than $1.1R_{t_2}C_{t_2}$, the parallel discharge of C_{t_2} by monostable A begins to have an effect on the output. This is illustrated by the timing diagram for retriggerable operation, which shows the voltage across C_{t_2} being returned to zero with each input pulse. As a consequence, C_{t_2} never reaches the 2/3V+ voltage threshold of timer B; thus, timer B never times out. Under these conditions, the output remains at a steady, high dc level. It will continue to do so as long as input triggers are received at a period less than $1.1R_{t_2}C_{t_2}$.

This circuit can be scaled to operate at virtually any rate consistent with the range of component values allowable with a 555. The output pulse width of the input section should be maintained at a small percentage of that of the second section. However, this time must be long enough to ensure complete discharge of C_{t_2} during the timing period of the input section. Other than this criterion, there is little that is critical regarding the input section.

4.9 DELAYED-PULSE GENERATION

There are many different monostable timer arrangements that fall under the general category of delayed-pulse generation. A few examples of these circuits are discussed in this section.

In Fig. 4-9, a circuit is shown that produces a pulse of width T_2 occurring at a time T_1 after an input trigger. This circuit uses a 556 timer to minimize the number of IC packages. The circuit is very straightforward and needs little explanation regarding its special features other than the input differentiator shown between sections, which will generally be necessary. Otherwise this circuit is a classic example of two cascaded, 555-type monostable timers.

Another example of a circuit that performs the function of delayed-pulse generation is shown in Fig. 4-10. This circuit uses individual 322 (or 3905) units, and although it requires more IC packages, it does permit direct interfacing between sections without differentiation. It also permits negative-going output pulses, a wider range of timing periods, and a wider range of supply voltages. This circuit uses the higher-speed characteristics of the 322 to advantage, generating a nominal 4.7-μs pulse after a delay of 10 μs. With timing periods of 10 μs or less, some trimming may be necessary with R_t or C_t because of limitations in predictability and propagation delays.

These concepts can be extended further by using the proceding circuits or others previously described to build general timing systems, as shown in Fig. 4-11. Here a "bank" of timers is shown that are triggered simultaneously by a single input. They produce three separate outputs of widths T_1, T_2, and T_3, respectively. The actual timer circuits used could be any described thus far for monostable operation. The example shown indicates positive triggering with negative output.

Similarly, a "chain" of timers can be constructed as in Fig. 4-12. Here, three delays of widths T_1, T_2, and T_3 are operated in sequence. Thus, T_1 triggers T_2; T_2 triggers T_3; and so on, for more delays. This type of system is also referred to as a *sequential* timer. It also could use any of the monostable timers described thus far.

4.10 PROGRAMMABLE MONOSTABLES

It is often necessary to be able to remotely select different time delays. The circuits of Figs. 4-13 and 4-14 illustrate methods of doing this with a minimum of circuit complexity and cost. In Fig. 4-13, a 555 is connected in a standard monostable arrangement, with the exception of R_t. Here, R_t is switch selectable by A_2, a 4016 CMOS quad analog switch. Two sections of this switch, S_1 and S_2, are used

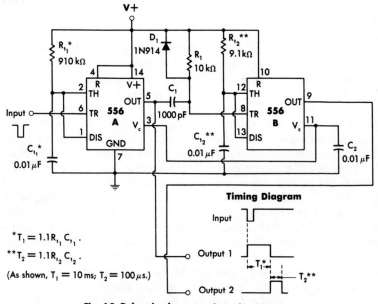

$$^*T_1 = 1.1 R_{t_1} C_{t_1}.$$
$$^{**}T_2 = 1.1 R_{t_2} C_{t_2}.$$
(As shown, $T_1 = 10$ ms; $T_2 = 100 \mu$s.)

Fig. 4-9. Delayed pulse generation using 556 type.

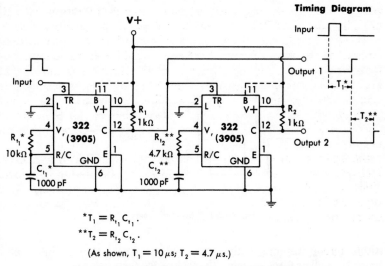

$$*T_1 = R_{t_1}C_{t_1}.$$
$$**T_2 = R_{t_2}C_{t_2}.$$

(As shown, $T_1 = 10\,\mu s$; $T_2 = 4.7\,\mu s$.)

Fig. 4-10. Delayed pulse generation using 322 (3905) type.

to select R_{t_1} or R_{t_2}. Section S_3 is used to invert the input control signal so that the S_1–S_2 switch states are reversed.

When the control line is high, S_1 is "on," connecting R_{t_1} to V+. The monostable timing period is then $1.1R_{t_1}C_t$. When the control line is low, S_2 is "on," connecting R_{t_2} to V+. The timing period then is $1.1R_{t_2}C_t$. Resistors R_{t_1} and R_{t_2} can be any two resistances within the capability of the 555.

Although this switching arrangement is shown within a basic 555 monostable circuit, it may also be applied to virtually any mono-

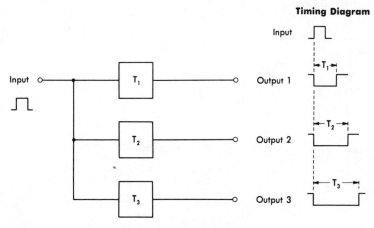

Fig. 4-11. A "bank" of timers.

99

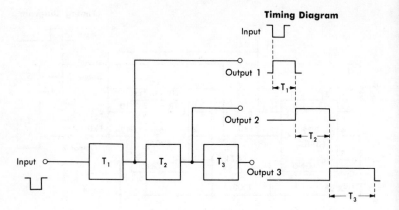

Fig. 4-12. A "chain" of timers.

stable circuit discussed thus far. This includes the modified 555 types, as well as circuits using the 322 and 3905.

For remotely selectable monostable timing periods that are integral multiples of one another, a programmable timer/counter is an effective performer, as shown in Fig. 4-14. This circuit is in essence a basic 2240 binary (or 2250 for bcd) programmable monostable

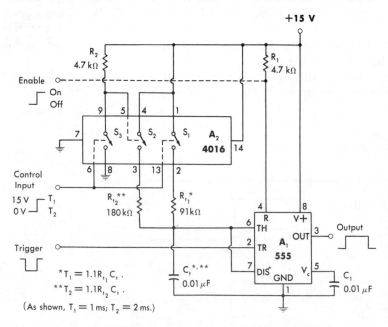

Fig. 4-13. Selectable-time programmable monostable using 555 type.

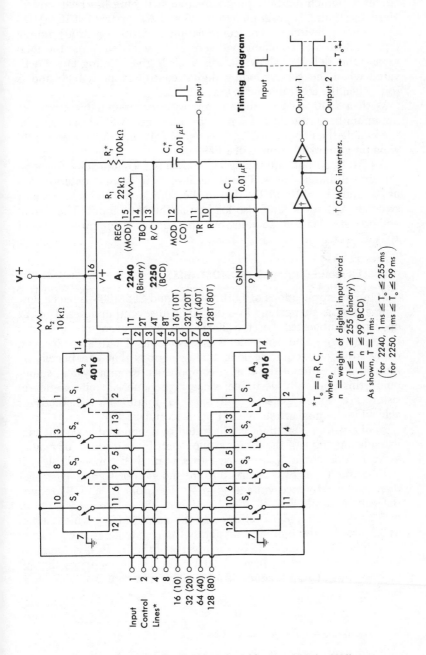

Fig. 4-14. Binary (or bcd) programmable monostable using 2240 (or 2250) type.

circuit, to which digitally programmable switching has been added. Here, the timing is programmed by A_2 and A_3, a pair of 4016 CMOS quad analog switches. The combination of 2240 (or 2250) timing taps connected to common load resistor R_2 will determine the total monostable delay, as shown in Fig. 4-14. A given timing tap is activated when the corresponding digital control input is high and is "off" when the control input is low.

With a 2240 used as shown, timing is programmable from the minimum basic interval of 1 ms, up to 255 ms. A 2250 may also be used (note the minor pin differences at pins 12 and 15), which would yield programmable timing of 1 to 99 ms.

A CMOS output buffer stage as shown is recommended for completely valid logic levels. This is because there is some deterioration of "0" and "1" states at R_2, due to the additional "on" resistance of switches A_2 and A_3, and also loading at the reset input. The circuit will operate over 5 to 15 volts, but operation is optimum at a V+ of 10 to 15 volts.

4.11 EXTENDED-RANGE MONOSTABLES

Standard applications of both the 555 and 322/3905 timers allow timing periods that extend into the range of several minutes. The 322 operating unboosted (or the 3905) can reach even into the hour range, if high-value components are available. There are, however, many situations requiring long timing periods that can benefit by reduction in value of the timing components. From a system standpoint, this is usually desirable, since high-value capacitors and resistors are both expensive and hard to obtain. Two methods of achieving this goal are shown in Figs. 4-15 and 4-16.

In Fig. 4-15, a 322 timer is illustrated in a monostable circuit that extends the range of the timing components by two orders of magnitude. In this circuit (as opposed to a normal 322 monostable), it will be noted that R_t is returned to the junction of R_1 and R_2 rather than to the reference voltage (pin 4). Resistors R_1 and R_2 reduce the voltage applied to R_t to a fraction of the reference voltage of 3.15 volts. Resistor R_2 is returned to the output of A_2, which serves as a buffer for the timing ramp across C_t. This technique effectively "multiplies" the value of R_t by the inverse of the R_1–R_2 division ratio. In this case R_1 and R_2 have a 100:1 ratio, which increases the apparent value of R_t 101 times. The resulting new timing expression is then:

$$T = \frac{R_1 + R_2}{R_2}\,(R_tC_t),$$

or

$$T = 101R_tC_t.$$

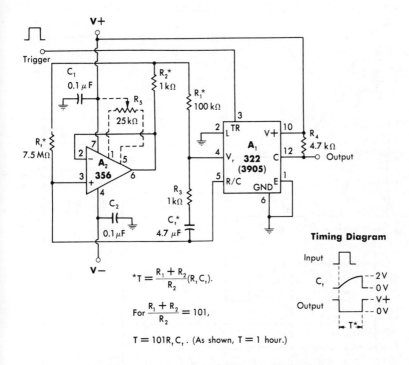

$$*T = \frac{R_1 + R_2}{R_2}(R_t C_t).$$

For $\frac{R_1 + R_2}{R_2} = 101,$

$T = 101 R_t C_t$. (As shown, $T = 1$ hour.)

Timing Diagram

Fig. 4-15. Extended-range monostable using 322 (3905) type.

In the example shown, an R_t of 7.5 MΩ and a C_t of 4.7 μF achieve a timing period of 1 hour. For such long periods, it is desirable to defeat the time-out-on-startup characteristic of the 322. This is done by returning C_t to the reference voltage instead of to ground.

The penalty extracted for this extended component-range capability is the addition of amplifier A_2, a FET-input device. This device must have a low input current; thus the 356 is chosen. Also, it should be stable in terms of input voltage, since the reference voltage in this circuit is only 31 mV. Calibration of the timing period can be accomplished either by trimming R_2 or by using the offset control, R5. The circuit uses dual supplies, which can range from ±5 volts to ±15 volts. Operation is virtually immune to supply-voltage variations because of the internal regulation of the 322.

A circuit that operates similarly using the 555 is shown in Fig. 4-16. A buffer amplifier is again used, in this case a 3140. Pin 7 of the 555, which would normally be used to clamp C_t, is freed for use as a second output in the circuit to avoid its leakage, which would otherwise fix the low timing-current limit. Its function is handled by a 4016 CMOS switch, which by contrast has a 100-pA leakage.

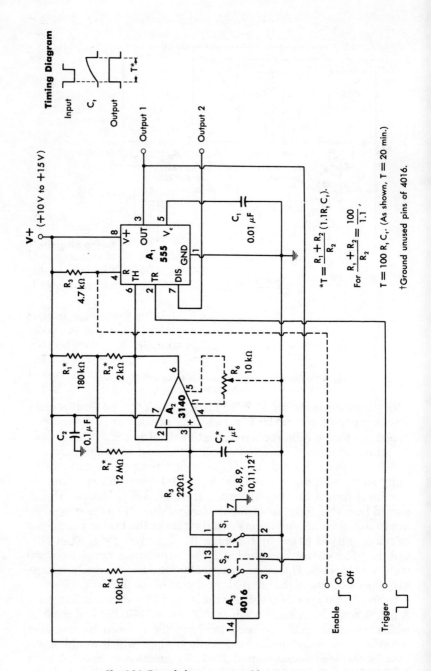

Fig. 4-16. Extended-range monostable using 555 type.

The V+ voltage is divided by R_1 and R_2 prior to being applied to R_t. In this case, R_1 and R_2 yield a division ratio of 90:1. This, when multiplied by the basic 555 timing period of $1.1R_tC_t$, achieves a new timing period of $T = 100R_tC_t$.

This circuit is most useful with single supplies in the range of +10 volts to +15 volts. Timing can be trimmed with R_2, or with optional control R_6. Although a timing interval of 20 minutes is shown, it can range much longer if desired. This circuit is somewhat more flexible than the 322 version since it has two outputs, pin 7 of the 555 having been made available as an uncommitted second output. It also has an enable input, if one is desired.

4.12 VOLTAGE-CONTROLLED MONOSTABLES

A useful option available with both the 555 and 322 timers is the ability to control their output pulse width with a variable voltage. This feature may be used to either trim the pulse width to a desired time interval, or to control it from an externally derived voltage for voltage-controlled (or modulated) applications. This technique is illustrated in Fig. 4-17.

In Fig. 4-17A, a 555 monostable circuit is shown with a control voltage applied to pin 5, the voltage-control pin. To trim the pulse width to a desired time, a 10-kΩ potentiometer, R_{adj}, can be connected across the V+ line as shown, with its arm going to pin 5. Variation of the potentiometer output will then vary the timing period of the monostable (over a fairly broad range, if desired). With the potentiometer connected in this fashion, operation will continue to be supply-voltage independent.

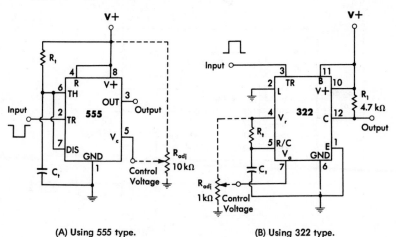

(A) Using 555 type.   (B) Using 322 type.

Fig. 4-17. Voltage-controlled monostable circuits.

An external voltage can also be applied to pin 5 (in lieu of the potentiometer) and will vary the timing period also. This can be used for pulse-width modulation, for instance. Operable limits are from approximately 12 volts down to 2 volts at a V+ of 15 volts, and proportionately less with lower supply voltages. The pulse-width-versus-voltage-control characteristic is not linear.

A 322 timer can also be operated in a voltage-controlled configuration, as shown in Fig. 4-17B. Here the 3.15-volt internal reference voltage is applied to a 1-kΩ potentiometer, R_{adj}, with the arm applied to the control-voltage terminal, pin 7. Operation is generally similar to that described for the 555, but a greater dynamic range of control is possible. The timing may be adjusted about the nominal 2-volt setting for trim purposes or may be taken all the way to zero for extreme variation in pulse width. The dynamic range of adjustment capability of this circuit approaches 100:1 or more. Furthermore, due to the internal regulator, it is highly independent of supply voltage.

Although operation is basically nonlinear, it can be made approximately linear if the pin 7 voltage is a small fraction of 2 volts; that is to say, very low on the R/C charging curve. External voltages can also be applied to this circuit for pulse-width modulation, if desired.

A 2240 programmable timer/counter may also be voltage controlled, as shown in Fig. 4-18. The time-base section of this device is somewhat similar to that of a 555 in operation, with the voltage-control terminal (pin 12) nominally at 2/3V+. It is adjusted or mod-

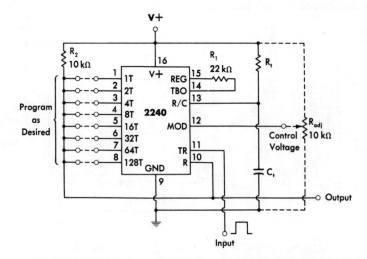

Fig. 4-18. Programmable voltage-controlled monostable using 2240 type.

ulated in a manner similar to a 555. Similar comments apply to other members of the timer/counter family.

The safe limits of the control-voltage pins must be observed with any of the devices used in voltage-controlled circuits. Care should be taken in lead length and dress, as extraneous noise introduced into a control terminal can induce undesirable jitter on the output pulse.

4.13 LINEAR-RAMP MONOSTABLES

A very useful modification to the standard monostable configuration is to make the timing ramp a *linear* waveform, rather than an exponential one. This not only makes the waveform much more useful externally, but also simplifies the timing and opens up new application possibilities.

One of the simplest ways to implement a linear-ramp monostable is to replace the timing resistor, R_t, with a constant-current source. This is shown within a 555 monostable circuit in Fig. 4-19. Here, Q_1 is the current-source transistor, supplying constant current I_t to timing capacitor C_t. When the timer is triggered, the clamp on C_t is removed and C_t charges linearly toward V+ by virtue of the constant current supplied by Q_1. Without R_3, the threshold at pin 6

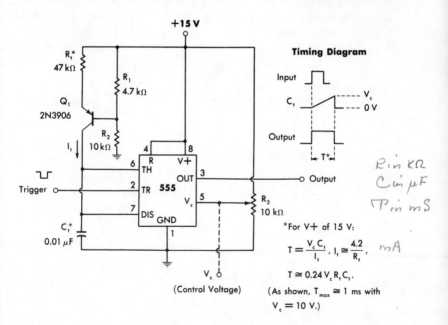

R_t in KΩ
C_t in μF
T_m in mS

*For V+ of 15 V:

$$T = \frac{V_c C_t}{I_t}, \quad I_t \cong \frac{4.2}{R_t}, \quad mA$$

$$T \cong 0.24 V_c R_t C_t.$$

(As shown, $T_{max} \cong 1$ ms with $V_c = 10$ V.)

Fig. 4-19. Linear-ramp monostable using 555 type.

would be 2/3V+; here, it is generally termed V_c. When the C_t voltage reaches V_c volts, the timing cycle ends. The general timing expression for output pulse T is

$$T = \frac{V_c C_t}{I_t},$$

where V_c is the voltage at pin 5, and I_t is the current supplied by Q_1. Current I_t can be approximated for a V+ of 15 volts as

$$I_t \cong \frac{4.2}{R_t}.$$

Then T is

$$T \cong 0.24 V_c R_t C_t.$$

Note that this expression is not exact. As a particular example, if $V_c = 10$ volts, $R_t = 47$ kΩ, and $C_t = 0.01$ μF, T works out to be slightly over 1 ms. Current I_t is, of course, linearly variable by R_3 or by an external voltage applied to pin 5.

In general, I_t should be 1 mA or less, and C_t can be any value compatible with the 555. The dynamic range of V_c in this circuit is 4:1 or more, operating at 15 volts. The circuit will function at lower voltages, but with some sacrifice in stability and accuracy.

Another simple, linear-ramp monostable circuit is shown in Fig. 4-20. This circuit uses a 322 as the timing device. In this circuit, Q_1 is the constant-current source, developing a timing current, I_t. Transistor Q_1 provides a constant current that is effectively regulated

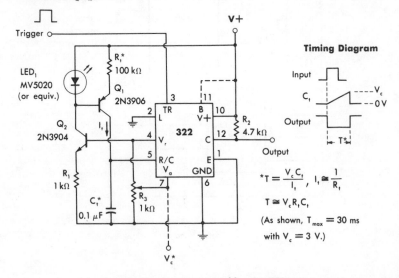

Fig. 4-20. Linear-ramp monostable using 322 type.

by light-emitting diode LED_1. The typical LED forward voltage drop of 1.6 volts is applied to Q_1, causing about 1 volt to be dropped across R_t. The timing current flowing in Q_1 is then

$$I_t \cong \frac{1}{R_t}.$$

The general timing expression for output pulse T is

$$T = \frac{V_c C_t}{I_t}.$$

With substitution for I_t, this becomes

$$T \cong V_c R_t C_t.$$

The inaccuracy in this equation is due to the relatively poor predictability of the LED and Q_1 voltage drops.

The LED is biased by a second constant-current source, Q_2. Transistor Q_2 is biased by the 3.15-volt reference voltage of the 322, and sets up a current of 3 mA in the LED. The use of the internal reference voltage to bias the LED gives the circuit a high immunity to supply-voltage variations. If this is not necessary, Q_2 may be deleted and Q_1 biased with a simple 4.7-kΩ resistor from base to ground. As shown, the circuit is linearly voltage controllable by R_3 (or V_c, if external) when V_c is 3 volts. For best performance, I_t should be kept below 100 μA.

4.14 FAST VOLTAGE-TO-PULSE-WIDTH CONVERTER*

Fig. 4-21 shows a circuit that is basically the same in concept as the simple linear-ramp monostables, but that uses different devices to accomplish the same function and thus offers several improvements in operation. In this circuit, a 322 is used as the monostable timer for several reasons. It has a higher speed capability, a wider range of voltage control, and better linearity.

The current source (A_2, an LM317) is three-terminal voltage regulator with an output voltage of 1.25 volts, which appears across the OUT and ADJ (or common) terminals. It will maintain this voltage accurately in spite of input-voltage fluctuations, loading at the output terminal, or a varying voltage at the ADJ terminal. Thus, A_2 can be used as both a current source and a voltage buffer.

The current, I_t, is set up by R_t and the bias current of A_2 according to the expression

---

* W. G. Jung, "Power Ramp Generator Delivers an Easily Adjustable 1A Output," *Electronic Design*, March 1, 1976.

$$I_t = 50 \ \mu A + \left(\frac{1.25}{R_t}\right).$$

This constant current charges C_t, forming a linear ramp of voltage at pin 5 of the 322. The same voltage also appears at the ramp output, shifted upward 1.25 volts (the output voltage of the 317). Loading at the ramp output does not affect the timing.

The high performance of both the 317 and the 322 enables this circuit to be very stable and highly immune to supply-voltage fluctuations. The width of the output pulse can be varied over a range of about 100:1, from a V_c of 3 volts downward. Due to the speed capability of both devices, operation down to a few microseconds of output pulse width is possible.

In the example of Fig. 4-21, the conversion sensitivity is 100 μs per volt of V_c. Conversion sensitivity is set by R_t, which trims I_t. It is best to maintain I_t within the range of 100 μA to 1 mA. (Note that part of I_t is due to the 50-μA bias current from the ADJ terminal of the 317.)

4.15 LONG-PERIOD VOLTAGE-CONTROLLED TIMER

The circuit of Fig. 4-22 uses an extension of the principle set forth in Section 4.14 to create a timer that features an even wider range of operation, in addition to the basic feature of linear voltage control.

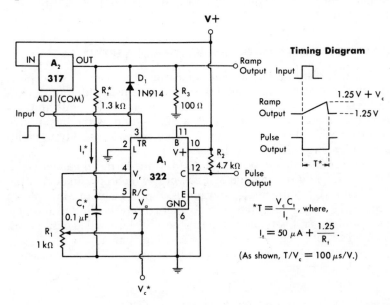

Fig. 4-21. Fast voltage-to-pulse-width converter.

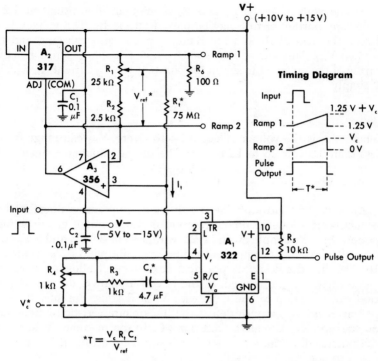

$$*T = \frac{V_c R_t C_t}{V_{ref}}$$

(As shown, T = 1 hr with V_c = 3 V and V_{ref} calibrated for 0.3 V.)

Fig. 4-22. Long-period voltage-controlled timer.

This circuit is easily calibrated and adjusted, and has a variable timing range of up to one hour.

In this circuit, a 322 is again used as the basic timing device, but with the constant current for timing supplied by a current source consisting of A_2 and A_3. Amplifier A_3 is a FET-input op amp, used here as a buffer to isolate the high-impedance R/C timing node. The typical 30-pA input current of this device ensures that the major bias-current error of the circuit is that of the 322 (unboosted). This allows I_t to range downward to about 1 nA for predictable operation.

Since A_3 is connected as a voltage follower, it "bootstraps" the timing voltage created across C_t. A replica of this linear voltage ramp appears as Ramp 2. During the timing interval, Ramp 2 will sweep from zero volts up to V_c (the control voltage), which is applied either via R_4 or from an external source.

The 1.25-volt three-terminal regulator, A_2, "floats" atop the Ramp 2 voltage. It maintains its 1.25-volt output across R_1 and R_2, and thus also delivers a level-shifted output, Ramp 1. Ramp 1 will sweep from 1.25 volts up to 1.25 volts plus V_c during the timing cycle.

111

As the voltage across R_1 and R_2 is maintained constant at 1.25 volts, any portion of this may be picked off via R_1. The voltage between the arm of R_1 and the Ramp 2 bus is also constant, and is termed V_{ref}. This low and constant voltage is applied across R_t (by virtue of the op amp) to set up the constant timing current, I_t, which is simply

$$I_t = \frac{V_{ref}}{R_t},$$

where V_{ref} is the voltage selected by R_1. Here, V_{ref} can range from less than 0.3 volt up to 1.25 volts. The timing equation is then

$$T = \frac{V_c R_t C_t}{V_{ref}}.$$

An important virtue of the use of a low reference voltage is the reduction it allows in timing-component values for a given timing period. In the example of Fig. 4-22, with a V_c of 3 volts and a V_{ref} of 0.3 volt, the timing-component values are multiplied by a factor of 10. This has the desirable practical advantage of allowing smaller-value capacitors to be used. In the circuit as shown, the timing period can range up to one hour with V_c at 3 volts. For such a long period, it is necessary to defeat the 322's characteristic of time-out on startup. This is done by returning C_t to the reference voltage of the timer, rather than to ground. Series resistor R_3 limits the charging current of C_t, for reliability reasons.

This circuit can easily be calibrated for selection of timing intervals by using a dial on R_4 or by using a switched voltage divider. Triggering can be accomplished with a simple push-button switch wired to the timer's reference voltage, since switch bounce will not interfere with such long timing periods.

It should be noted that this circuit requires a negative supply voltage for A_3. The supply need not be regulated, however, and can be in the range of -5 to -15 volts. The positive supply is also noncritical. Both supplies should be bypassed for high frequencies, preferably close to A_3.

4.16 RATIOMETRIC VOLTAGE-TO-PULSE-WIDTH CONVERTER

Another circuit that is useful as a voltage-to-pulse-width converter, but that has additional features of operation, is shown in Fig. 4-23. This version does not have a buffered, linear-ramp output, but it does have the inherent property of cancelling any changes in the reference voltage, and thus offers almost total immunity to supply-voltage-induced timing changes.

In the circuit of Fig. 4-23, A_1 is a 322 connected as a linear voltage-to-pulse-width controller. A linear timing ramp appears across C_t, and the pulse width is linearly variable via V_c. The feature of the circuit that makes it ratiometric is the type of constant-current source used. Here, Q_1 sets up a current through R_2. This current will be equal to the 322 reference voltage divided by R_t, since the V_{BE} of Q_1 and the V_{BE} of Q_2 cancel each other. The op amp, A_2, causes the current in R_2 to be duplicated in R_3. The current in Q_2 is then used as the charging current for C_t. Various values of R_t can be connected from the emitter of Q_1 to ground, and will program a corresponding

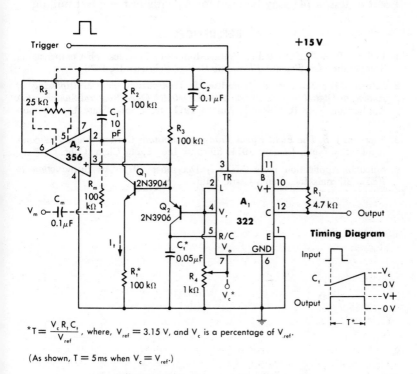

$$*T = \frac{V_c R_t C_t}{V_{ref}}, \text{ where, } V_{ref} = 3.15 \text{ V, and } V_c \text{ is a percentage of } V_{ref}.$$

(As shown, T = 5 ms when $V_c = V_{ref}$.)

Fig. 4-23. Ratiometric voltage-to-pulse-width converter.

timing current in C_t over a wide range. Values for R_t can span two to three decades ranging from a minimum of 50 kΩ upward. However, if a three-decade range of operation is desired, the offset voltage of A_2 should be nulled with R_5, an optional trim network, and the boost connection on the 322 (pin 11) should be deleted.

The circuit is ratiometric because with a given timing current set up in C_t, which is proportional to the reference voltage of the 322, a percentage of this reference voltage is used as V_c. Therefore, changes in V_{ref} cancel out, in a manner similar to the basic timing of the 322 as an exponential ramp timer.

The circuit is quite flexible, and can be set up for a wide range of operational modes other than that shown. For example, it may be used as a linear pulse-width modulator by applying an ac signal to the optional V_m point as shown. R_m is a series resistor used to scale the peak modulation current so that it is equal to I_t; it should be equal in value to R_t with peak modulation voltages of 3.15 volts. Capacitor C_m is an ac coupling capacitor chosen for the desired 3-dB rolloff point (16 Hz as shown). If modulation is not used, and R_t is 1 MΩ or less, a 741 may be used for A_2 without the offset nulling.

REFERENCES

1. Lefferts, P. A. "LED Used as Voltage Reference Provides Self-Compensating Temperature Coefficient." *Electronic Design,* February 15, 1975.

2. Litus, J. Jr.; Niemiec, S.; Paradise, J. *Transmission and Multiplexing of Analog or Digital Signals Utilizing the CD4016A Quad Bilateral Switch.* RCA Application Note ICAN-6601, August 1971. RCA Solid State Div., Somerville, N.J.

3. Sherwin, J. S. *The Field Effect Transistor Constant Current Source.* Siliconix Application Note, January 1971. Siliconix Inc., Santa Clara, Calif.

4. Siliconix Application Note AN73-7, December 1973. *An Introduction to FETs.* Siliconix Inc., Santa Clara, Calif.

5

Astable Timer Circuits

In this chapter, various modified astable configurations using different timer devices are discussed. As noted previously in Chapter 4, many of these circuits will be seen applied within the general applications section (Chapter 6) as portions of larger-scale systems. A broad variety of circuits are shown, and many of them can be modified further should the reader desire to do so. Equations governing circuit operation are given, limits of performance are discussed, and suggestions for modifications are presented.

5.1 MINIMUM-COMPONENT ASTABLE*

This circuit (Fig. 5-1) is so named because of its obvious simplicity. There are only two actual frequency-determining components—R_t and C_t. The noise bypass, C_1, may even be omitted when it is not necessary, making the circuit a 3-component astable.

In the standard 555 astable, the junction of the two timing resistors is switched to ground by pin 7. Two of the resistors and C_t determine the output high time; one of them and C_t determine the low time. In theory, identical times can be generated if a single timing resistor is switched between the charging voltage and ground, with the upper and lower thresholds fixed. It is on this theory that this astable is based.

The circuit generates an approximately square output waveform as R_t is switched by the output from ground to near V+. Ideally, the output voltage would be V+ in the high state; however, the

---

* S. A. Orrel, "IC Timer Plus Resistor Can Produce Square Waves," *Electronics,* June 21, 1973; Copyright © McGraw-Hill, Inc., 1973.

high-state saturation limit of a 555 is nearly 2 volts below V+. This causes some time asymmetry in the waveform, particularly at low supply voltages near 5 volts. The timing period would be simply 2 × 0.693R_tC_t (or 1.386R_tC_t) were it not for this error. In light of this error, the time equation given is approximately 1.4R_tC_t, and this will vary with differing supply-voltage levels. However, even in view of this limitation, the circuit is still quite useful due to its inherent simplicity and low cost.

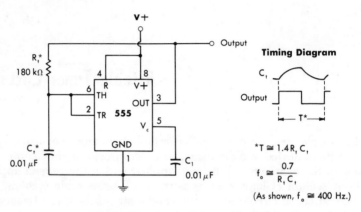

Fig. 5-1. Minimum-component astable circuit.

5.2 IMPROVED MINIMUM-COMPONENT ASTABLE

With a slight modification, the minimum-component astable of Section 5.1 can be improved as shown in Fig. 5-2. Here R_1 is an added pull-up resistor, which forces the pin 3 voltage to rise to V+ in the high state. This removes the time asymmetry error, and the waveform becomes square in shape. The time (and frequency) expressions then become more precise, as shown. This assumes that the loading at pin 3 for the high state is low.

If desired, an additional (optional) output is available at pin 7, with timing identical to pin 3. This output may be referred to any supply voltage from zero to +15 volts, regardless of the timer's actual supply voltage; thus it can be readily compatible with either TTL or CMOS logic.

5.3 SQUARE-WAVE ASTABLES*

The square-wave astables shown in Figs. 5-3 and 5-4 are similar to the two previously described, but they offer further improvement in precision. The circuit of Fig. 5-3 is similar to the basic 555 astable

---

* W. G. Jung, "The IC Time Machine," *Popular Electronics*, January 1974.

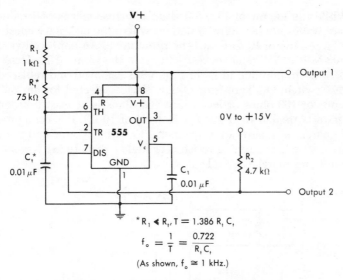

Fig. 5-2. Improved minimum-component astable circuit.

except for the fact that R_1 is made a very small fraction of R_t. This will provide a nearly equal charge and discharge resistance path for C_t. As a result, the timing periods are equal, or at least equal consistent with the match of resistances. In practice, the output very closely approximates a square wave.

A virtue of this configuration is that the timing network is not driven from the output; therefore, loading at pin 3 will have no effect on timing. Pin 3 may then be loaded on a relatively unrestricted basis.

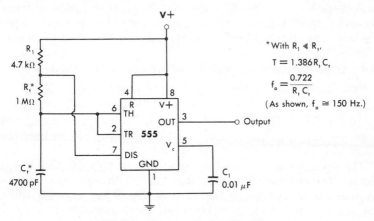

Fig. 5-3. Basic square-wave astable circuit.

While the circuit of Fig. 5-3 is an effective means of producing square waves, there will be situations where the necessary constraint of a large ratio of R_t to R_1 will be prohibitive. An improved version, shown in Fig. 5-4, removes restrictions on the value of R_t. Here, the resistor corresponding to R_1 of Fig. 5-3 is removed and replaced by a FET switch, Q_1. Transistor Q_1 is a device selected for an "on" resistance of 100 ohms or less. Since 100 ohms is at least 2 orders of magnitude smaller than a lower R_t limit of 10 kΩ, this configuration will guarantee square-wave symmetry to within 1% or better, down to $R_t = 10$ kΩ. For timing resistances higher than 10 kΩ, asymmetry due to R_t mismatch is even less.

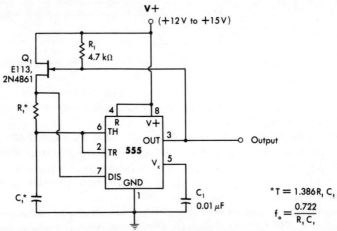

Fig. 5-4. Improved square-wave astable circuit.

Transistor Q_1 is driven from the output of the timer, which must pull up to V+ to switch Q_1 fully "on." Thus, R_1 is added to ensure this. The circuit will work over the full range of frequencies of which the 555 is capable, and will perform best at a V+ of 12 to 15 volts.

5.4 SQUARE-WAVE ASTABLES WITH EXTENDED RANGE

The timing range of the square-wave astable circuit of Fig. 5-5 is extended far beyond that of the basic 555 limits, by the use of an amplifier that buffers the timing network. This amplifier completely removes the input bias-current restrictions of the 555 and replaces them with those of the op amp used for A_2.

The op amp shown (a 3130) is a single-supply MOSFET-input device. The extremely low bias current of this device (5 pA) removes virtually all restrictions on timing error due to bias currents, allowing timing resistances up to hundreds of megohms to be used. In the example shown, the 260-MΩ resistor and 10-µF capacitor yield

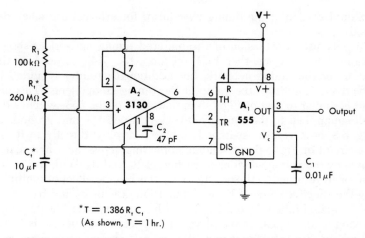

*$T = 1.386\,R_t\,C_t$
(As shown, $T = 1$ hr.)

Fig. 5-5. Square-wave astable with range extended by buffering the timing network.

a 1-hour timing interval. This is not the upper limit of the circuit by any means, as R_t can be much higher if the leakage resistance of C_t and the surrounding circuit is also appreciably higher.

The use of a buffer stage as A_2 is applicable to other timer circuits as well, either to extend the range of R_t and C_t or to provide a

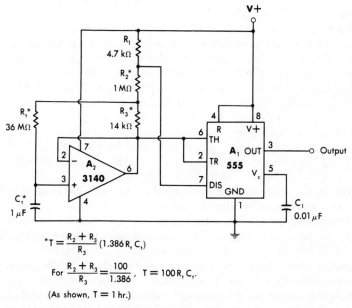

*$T = \dfrac{R_2 + R_3}{R_3}\,(1.386\,R_t\,C_t)$

For $\dfrac{R_2 + R_3}{R_3} = \dfrac{100}{1.386}$, $T = 100\,R_t\,C_t$.

(As shown, $T = 1$ hr.)

Fig. 5-6. Square-wave astable with range extended by buffering and by
resistance multiplication.

buffered version of the timing waveforms for external use when desired.

A very useful extension of the buffered timing network is shown in Fig. 5-6. Here, another FET op amp, the 3140, is used as the buffer for the timing network, but additional circuitry is added to effectively "multiply" the value of the timing components. This circuit is similar to that of Fig. 5-5 in that pin 7 of the 555 switches the voltage applied to the timing network between V+ and ground. In Fig. 5-5, however, the switched voltage is applied directly to R_t. In the circuit of Fig. 5-6, the voltage appearing across R_t is first scaled down by the voltage divider consisting of R_2 and R_3. With the "cold" end of R_3 returned to the output of A_2, this has the effect of multiplying the effective value of R_t by the division ratio of R_2 and R_3.

The actual ratio selected for R_2 and R_3 can vary over a wide range, as long as the square wave of voltage appearing across R_3 is about 50 mV or more. In this circuit, the ratio is chosen to be 100 times the reciprocal of twice the natural log of 2. This makes the timing expression for the astable simply

$$T = 100R_tC_t.$$

If desired, R_3 can be used to trim the timing period. The output waveform will remain essentially square, as long as R_2 is maintained much larger than R_1. This circuit is also supply independent in operation, as is the basic 555 astable.

An even further extension of the idea of "multiplying" the timing component values is shown in Fig. 5-7. Whereas the circuit of Fig. 5-6 multiplied the *resistance* value, this circuit multiplies the *capacitance* value. With the exception of op amp A_{2B}, this circuit is identical in concept to the basic buffered square-wave astable of Fig. 5-5. Note, however, that C_t is not returned to ground in this circuit but to the output of op amp A_{2B}, which is an amplifier that functions as a capacitance multiplier. The basic timing waveform, which appears at the output of op amp A_{2A}, is amplified by the ratio of R_3 to R_2 and fed back to the bottom end of C_t. This increases the effective value of C_t by the ratio of the gain of the A_{2B} stage, or R_3/R_2. In this example, the value of C_t is multiplied by 4.7, making it effectively a 4.7-μF capacitor rather than a 1-μF one.

The penalty for this "electronic gain" in capacitance is the addition of the amplifier stage A_{2B} and the necessity of a negative supply. The negative supply is not inherently necessary for operation, but for practical gain values, it is. To obtain useful gains in capacitance, the A_{2B} stage should amplify by a factor of 2 or more. To do this, it must be capable of swinging large amplitudes at its output, since its input is V+/3 volts, as set by the 555. In the case here, the gain of 4.7 causes A_{2B} to develop over a 20-volt p-p output at a V+ of 15 volts.

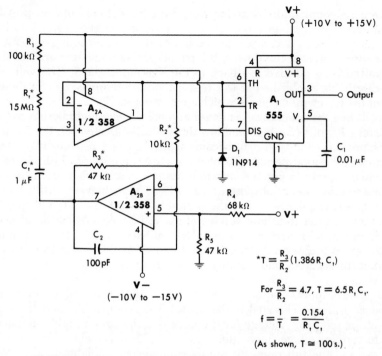

Fig. 5-7. Square-wave astable with range extended by buffering and by capacitance multiplication.

For this reason the V— supply should be equal and opposite the V+ supply.

The circuit is supply independent in operation, and works well over a 10- to 15-volt range. Other op-amp types may be used instead of the 358 type shown. For example, a FET type could be used for the voltage follower. The amplifier stage is not critical in terms of amplifier type, except from the standpoint of output swing capability. The 358 or 324 types are noteworthy in terms of this parameter.

It would seem that the marriage of resistance and capacitance multiplication techniques just described should be capable of yielding effective $R_t C_t$ multipliers of several hundred. The proof of this theory is left as a classic "exercise for the reader."

5.5 TYPE 322 SQUARE-WAVE ASTABLE

Although not primarily intended to be operated in the astable mode, 322-type timers may be configured to operate in this mode, as is shown in Fig. 5-8. Here, the 322 is wired essentially as an inverting comparator by connecting the reference terminal to pins 2 and 3. This disables the discharge transistor and allows pin 5 to be freely used as a comparator input. Then, by connecting the R_tC_t network around the device, the circuit can be made to oscillate by introducing positive feedback to the V_{adj} input, pin 7. The voltage thresholds established at this terminal will determine the end limits of the RC charging curve, which appears across C_t.

When the comparison voltages at pin 7 are chosen to be 1.96 volts and 1.19 volts (for high and low thresholds, respectively), the timing expressions for t_1 and t_2 become $0.5R_tC_t$. Quite conveniently, this makes the total period (T) of the square wave simply $1.0R_tC_t$. The circuit shown in Fig. 5-8 accomplishes this in somewhat rudimentary fashion by the use of diodes D_1 and D_2. The main asset of this technique is simplicity. When the output is low, D_1 and D_2 pull pin 7 down to their threshold voltage of about 1.2 volts. When the output is high, pin 7 reverts to its normal potential of 2 volts. This nominally achieves the desired objective of 1.96-volt and 1.19-volt thresholds, but the accuracy and stability of the circuit are limited by the diode voltage.

The output is a square wave, with an amplitude equal to the reference voltage. This makes the output TTL compatible, and the use of the reference voltage for timing makes the circuit highly immune to supply-voltage fluctuations, as in a 322 monostable. The circuit can use the full range of timing components available to the 322, and judicious use of the boost terminal allows either high speed or

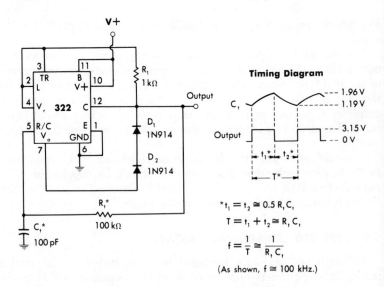

Fig. 5-8. Type 322 square-wave astable circuit.

very long timing-period operation. The boost terminal should be used above 1 kHz.

5.6 TYPE 2240 "GUARANTEED" SQUARE-WAVE ASTABLES

Under certain conditions, it may be necessary to fix the duty cycle of an astable timer precisely at 50%. The square-wave astables previously discussed approach inherently square-wave operation (some closely), but the accuracy and stability in some cases are subject to component tolerances and/or drift. The circuit of Fig. 5-9 generates an output waveform that is "guaranteed" to be a square wave, regardless of frequency.

This circuit takes advantage of the binary counter chain of a 2240. Since the output of a binary counter is inherently a square wave,

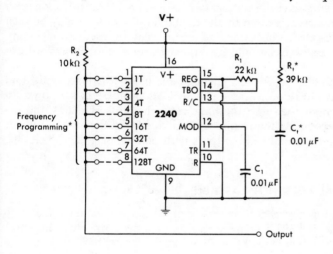

$$*f_o = \frac{1}{2nR_1C_1}, \text{ where,}$$

$n = 1, 2, 4, 8, 16, 32, 64,$ or $128.$

As shown:

n	f_o (Hz)
1	1280
2	640
4	320
8	160
16	80
32	40
64	20
128	10

Fig. 5-9. Type 2240 "guaranteed" square-wave astable circuit.

this ensures an output square wave for any frequency selected. Since the 2240 is also programmable, it actually yields an entire range of frequencies that can be programmed in binary fashion. If the time-base frequency is set to be a power of 2, the submultiple frequencies available with programming will fall as integer frequencies, as shown in the table of Fig. 5-9.

The circuit of Fig. 5-9 can also be made digitally programmable, in terms of output square-wave frequency, by substituting an analog switch for the simplified programming shown in Fig. 5-9. This circuit is shown in Fig. 5-10. Here, A_2 is an eight-input, one-output CMOS multiplexer—the 4051. This device has a three-line, channel-select control—inputs A, B, and C. These inputs are used to select one of the eight possible switching paths through the device. With all 0s for instance, input pin 13 is connected to output pin 3. As used here, this programs the output to a frequency of $\frac{1}{2}nR_tC_t$, where n is the number represented by the digital input word. As may be seen by the truth table, this three-line control may be used to select any one of the eight basic output frequencies of the 2240, and all are square-wave in form.

Output buffering is recommended for use with this circuit, and can take the form of a simple pair of CMOS inverters as shown. In the example of Fig. 5-10, the basic frequency is 1 Hz; thus, the period is 1 second. As the frequency is divided, the time is multiplied; thus, this circuit will yield outputs with periods of 1, 2, 4, 8, 16, 32, 64, or 128 seconds.

5.7 "DUAL ONE-SHOT" ASTABLE

Fig. 5-11 is an astable circuit that is actually made up of two cross-connected 322-type monostables, thus the term "dual one-shot" astable. The advantage of this circuit over a straight 555-type astable is in the flexibility and range of timing periods it offers; either t_1 or t_2 (or both) may be adjusted independently and without interaction. A second advantage, which can be important, is the availability of complemented outputs. A third advantage, although a relatively minor one, is the greater "mileage" extracted from the R_tC_t networks; the timing is $1.0R_tC_t$ rather then $0.693R_tC_t$—almost twice the delay for a given capacitor.

The circuit uses two 322s as shown, but it can also use two 3905s if the t_1 and t_2 timing periods are 1 ms or greater. The 322s have the optional boost feature, which allows timing periods shorter than 10 μs. Since the upper timing-interval limit is on the order of minutes (for standard R_tC_t values), the timing asymmetry can be as high as 10,000,000/1 with this circuit, with each timing interval being very predictable.

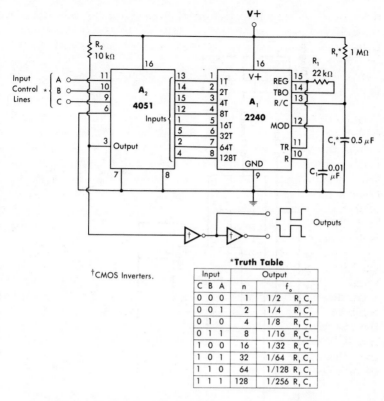

Fig. 5-10. Digitally programmable "guaranteed" square-wave astable circuit.

†CMOS Inverters.

*Truth Table

Input			Output	
C	B	A	n	f_o
0	0	0	1	1/2 $R_1 C_1$
0	0	1	2	1/4 $R_1 C_1$
0	1	0	4	1/8 $R_1 C_1$
0	1	1	8	1/16 $R_1 C_1$
1	0	0	16	1/32 $R_1 C_1$
1	0	1	32	1/64 $R_1 C_1$
1	1	0	64	1/128 $R_1 C_1$
1	1	1	128	1/256 $R_1 C_1$

Because these devices (322 or 3905) have the design characteristic of timing out upon start-up, this astable circuit will always self-start at the time of turn-on. If desired, the circuit can be gated "on" and "off" with an npn open-collector transistor switch to ground at either trigger terminal. The conduction of this switch will inhibit oscillations. When the switch opens, it will trigger the 322 and thus start synchronous oscillations. Suitable switches are other 322 or 3905 collector outputs, or TTL open-collector stages such as a 7405 or 7407.

5.8 "CHAIN" ASTABLE

An extension of the dual one-shot astable circuit is the "chain" astable of Fig. 5-12. This circuit merely adds another monostable within the oscillating loop or "chain," thus its name. The circuit, however, is not limited to three timers; it may be extended further, if desired.

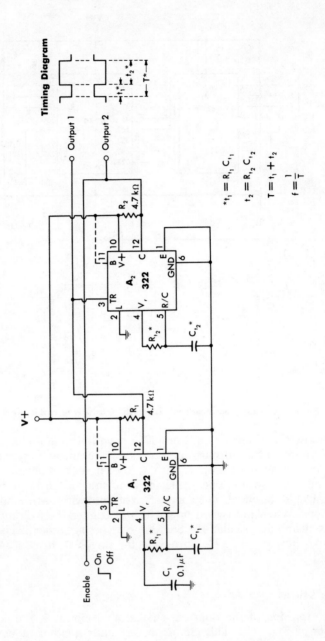

Fig. 5-11. "Dual one-shot" astable circuit.

126

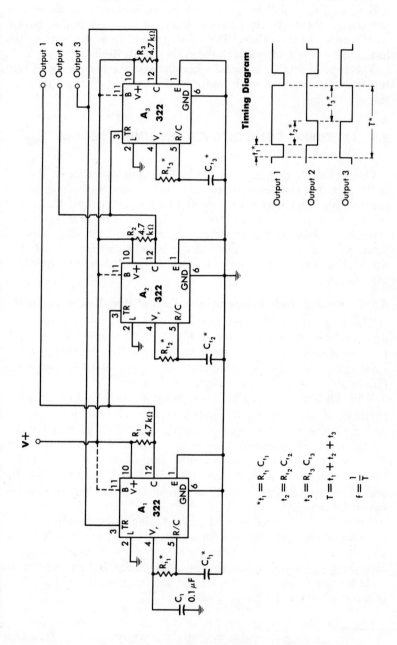

Fig. 5-12. "Chain" astable circuit.

$*t_1 = R_1 C_1$

$t_2 = R_{t_2} C_{t_2}$

$t_3 = R_{t_3} C_{t_3}$

$T = t_1 + t_2 + t_3$

$f = \dfrac{1}{T}$

With three timers within the loop, the timing periods are t_1, t_2, and t_3, and they will be generated in the sequence as shown. As previously, there are no inherent limits of any one timing period with respect to the others. The total period, T, is the sum of the individual periods, after which the cycle repeats itself.

This type of astable is useful where a prescribed sequence of timing events is required, with timing periods of an arbitrary relationship.

5.9 EXTENDING CONTROL OVER THE TIMING PERIODS OF THE 555

One of the disadvantages of the basic astable operation of the 555 is the fact that the timing periods, t_1 and t_2, are not independently controllable. This imposes a limitation on many designs that could otherwise make use of the device. There have been a number of schemes developed to circumvent these limitations, and some of them are most ingenious. Some of the better ideas are presented in this section, with references to the original sources where appropriate.

5.9.1 Astable With Independently Controllable Timing Periods*

The simplest modification to the basic 555 astable circuit is the addition of diode D_1, as shown in Fig. 5-13. This diode is a general-purpose silicon unit and is connected across R_{t_b}. Although on the surface this alteration seems quite simple, it does in fact give the 555 astable a much greater degree of freedom.

With D_1 added, C_t is charged through R_{t_a} and D_1 in series. The shunting of R_{t_b} with D_1 effectively removes R_{t_b} from the picture as a timing resistor during the charging cycle of C_t. During the discharge of C_t, D_1 is reverse biased, so this part of the circuit's function remains the same, with C_t discharging through R_{t_b} to form the period t_2. In effect, then, R_{t_a} and C_t control the period t_1, while R_{t_b} and C_t control the period t_2. There is little, if any, interaction between the timing periods.

Because of the presence of D_1, the t_1 timing expression is modified and can be approximated for a V+ of 15 volts as $t_1 = 0.76R_{t_a} C_t$. This expression varies according to the supply voltage and becomes less precise at lower voltages. Thus, the circuit is best used at 15 volts for this reason, and also for the reason of minimizing the temperature effects of the forward voltage of D_1.

---

* T. W. James, "Single Diode Extends Duty-Cycle Range of Astable Circuit Built With Timer IC," *Electronic Design*, March 1, 1973.

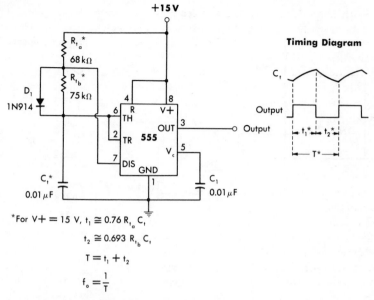

$$^*\text{For } V+ = 15 \text{ V}, \ t_1 \cong 0.76 \, R_{t_a} C_t$$

$$t_2 \cong 0.693 \, R_{t_b} C_t$$

$$T = t_1 + t_2$$

$$f_o = \frac{1}{T}$$

(As shown, t_1 and $t_2 \cong 500 \ \mu s$; $f_o \cong 1$ kHz.)

Fig. 5-13. Astable with independently controllable timing periods.

In the example shown, timing periods t_1 and t_2 are chosen to be equal at 500 μs, yielding a 1-kHz oscillation frequency. Other timing periods are readily realized, however, up to the extremes of timing component values usable with the 555.

5.9.2 Astable With Independently Adjustable Timing Periods*

The circuit of Fig. 5-14 is identical to that of Fig. 5-13, except for the addition of diode D_2 in series with R_{t_b}. This completely removes R_{t_b} from the charging path and makes the charge and discharge paths identical, which allows even greater independence of the timing periods, and a comparable ease of adjustment.

The equations for timing periods t_1 and t_2 in this circuit are the same: simply $t = 0.76 R_t C_t$, where R_t is R_{t_a} or R_{t_b}. Due to the presence of the diodes, both timing expressions will vary with the supply voltage; thus, the circuit should be operated at a V+ of 15 volts for best accuracy. Also, both timing periods are temperature sensitive because of the diodes, and this effect is minimized at 15 volts.

Resistors R_{t_a} and R_{t_b} can be adjusted over very wide ranges within the resistance limits of the 555. In the example shown, R_{t_a} is made

---

* M. S. Robbins, "IC Timer's Duty Cycle Can Stretch Over 99%," *Electronics*, June 21, 1973; Copyright © McGraw-Hill, Inc., 1973.

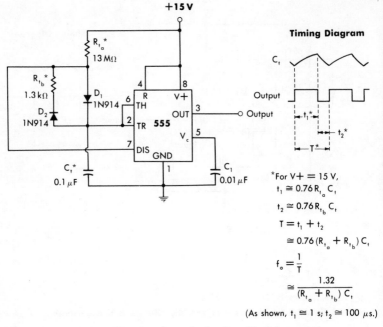

Fig. 5-14. Astable with independently adjustable timing periods.

10,000 times R_{t_b}, which yields a t_1/t_2 ratio of 10,000:1. With C_t equal to 0.1 μF, the circuit will yield a 100-μs output pulse once every second.

A very interesting version of this circuit results when the timing resistances, R_{t_a} and R_{t_b}, are made the center-to-end resistances of a single potentiometer. With the arm of the potentiometer centered, $R_{t_a} = R_{t_b}$, so the duty cycle will be 50%, producing output square waves. As the arm is varied to either side of center, R_{t_a} increases as R_{t_b} decreases (or vice-versa), but the total resistance remains the same. As a result, the duty cycle can be varied without changing the operating frequency with this type of adjustment.

5.9.3 Astable With Independently Adjustable Timing Periods and Auxiliary Output*

The circuit of Fig. 5-15 is very similar to that of Fig. 5-14, but it has one additional resistor, R_1. If desired, an auxiliary output is also available across this resistor; it is shown as Output 2. Loading must be kept light at this terminal, if used.

---

* A. R. Klinger, "Getting Extra Control Over Output Periods of IC Timer," *Electronics*, September 19, 1974; Copyright © McGraw-Hill, Inc., 1974.

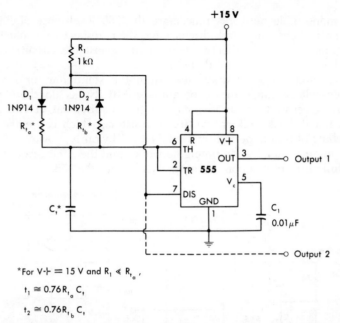

*For V+ = 15 V and $R_1 \ll R_{t_a}$,

$t_1 \cong 0.76R_{t_a}C_t$,

$t_2 \cong 0.76R_{t_b}C_t$,

Fig. 5-15. Astable with independently adjustable timing periods and auxiliary output.

The timing components (R_{t_a}, R_{t_b}, D_1, D_2, and C_t) perform functions similar to their counterparts in Fig. 5-14. If R_1 is made much lower than R_{t_a}, R_{t_a} will be dominant as the timing resistor. The timing equations are then simply $t = 0.76R_tC_t$, where R_t is R_{t_a} or R_{t_b}. It should be noted that any loading at the auxiliary output can influence the timing if it alters the high- or low-state voltage. Output 2 is therefore most suitable for CMOS or other minimal loads.

5.9.4 Astable With Separately Controllable Timing Periods*

An astable circuit that takes a completely different approach to the separation of the timing periods is shown in Fig. 5-16. This circuit is actually built around two types of 555 monostables which complement each other to form the timing periods, t_1 and t_2.

To the left of the circuit may be seen R_{t_a} and C_{t_a}, which are identical in hookup and operation to a standard 555 monostable. These two components control the output high period, $t_1 = 1.1R_{t_a}C_{t_a}$. To the right, R_{t_b} and C_{t_b}, in conjunction with D_1 and R_1, form an inverted 555 monostable with a timing period of $t_2 = 1.1R_{t_b}C_{t_b}$. These

---

* J. P. Carter, "Astable Operation of IC Timers Can Be Improved," *EDN*, June 20, 1973.

two monostable networks time-share the 555 level-sense circuitry, alternately charging and discharging their respective capacitors. The time relationships between the two monostable circuit halves may be noted in the timing diagram.

This circuit has no drawbacks from the standpoint of limited supply-voltage range or temperature sensitivity, as do those previously discussed. Its disadvantage is the requirement of two timing capacitors as well as two timing resistors, which can make the matching of timing periods more difficult. Obviously, however, there is no interaction between timing networks, and the circuit can utilize the full range of component values usable with the 555.

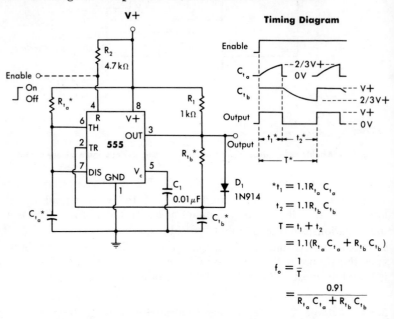

Fig. 5-16. Astable with separately controllable timing periods.

This circuit can easily be gated "on" and "off" by the use of the reset pin (pin 4) as shown. When this pin is low, the output is held low. When the enable line is taken high, the circuit starts oscillating synchronously with the t_1 period. The first cycle (and all succeeding cycles) are of the proper width, unlike the conventional 555 astable when gated.

5.9.5 Astable With Totally Independent and "Limitless" Timing Periods

The circuit of Fig. 5-17 approaches the problem of separating the timing periods of the 555 astable differently; it switches separate

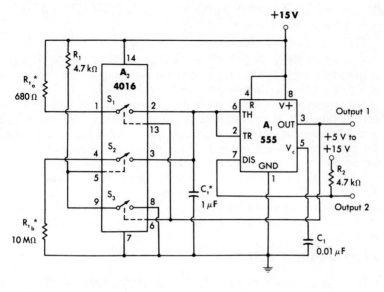

$$*t_1 = 0.693\, R_{t_a}\, C_t$$
$$t_2 = 0.693\, R_{t_b}\, C_t$$
$$T = t_1 + t_2$$
$$= 0.693\, (R_{t_a} + R_{t_b})\, C_t$$
$$f_o = \frac{1}{T}$$
$$= \frac{1.44}{(R_{t_a} + R_{t_b})\, C_t}$$

(As shown, $t_1 \cong 700\ \mu s$; $t_2 \cong 7$ s.)

Fig. 5-17. Astable with totally independent and "limitless" timing periods.

timing resistances in or out, according to the timing period that is active. The means to accomplish this is the CMOS analog switch, A_2. The device used here, a 4016, has a typical "on" resistance of 300 ohms and a typical off-state leakage of 100 pA. With this perform-ance, it sacrifices none of the basic input characteristics of the 555; yet it is a low-cost device.

When the output is high, or is in its t_1 state, S_1 is "on" and R_{t_a} is connected from V+ to C_t. The timing equation for t_1 is then simply $0.693R_{t_a}C_t$. When the output is low, S_1 is "off" and S_2 is "on." The output is inverted by switch section S_3, which causes the S_1 and S_2 on-off states to be opposite.

With S_2 "on," R_{t_a} is connected from C_t to ground. This causes the t_2 timing equation to be $0.693R_{t_b}C_t$. Both timing expressions are then of the form $t = 0.693 R_t C_t$. This also makes the total timing period the sum of t_1 and t_2, or $T = 0.693 (R_{t_a} + R_{t_b}) C_t$.

The advantage of this circuit is the total independence of R_{t_a} and R_{t_b}, which can range over the full gamut of component values usable with the 555. The only restriction occurs at timing resistances of a few kilohms or less, where the typical 300-ohm "on" resistance of the analog switch will become a significant part of R_t. In the example shown, t_1 is approximately 700 μs, while t_2 is 7 seconds, a dynamic range of 10,000:1. Although this is not actually "limitless," the practical limits are defined by the component values of the 555, not by the resistances of switch A_2. If desired, the range of resistance values can be extended even further by insertion of a buffer amplifier between C_t and the 555, in a manner similar to Fig. 5-5. A bonus of this circuit is the auxiliary output available at pin 7 of the timer, which can be referred to any voltage from zero to +15 volts.

5.10 SIMPLIFIED ASTABLE

In many applications, a precise and predictable duty cycle or timing ratio is *not* a requirement. However, a predictable and stable pulse rate (or frequency) may be. For this type of application, some simplifying assumptions can be made, which both ease the design process and provide additional circuit options.

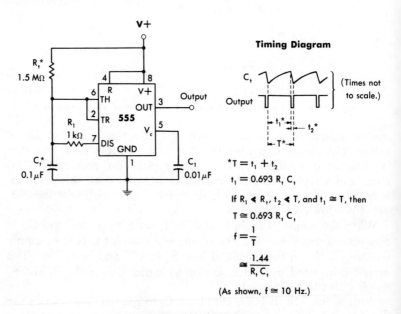

Fig. 5-18. Simplified 555 astable circuit.

In the circuit of Fig. 5-18, the basic 555 astable timing network is reduced to a single timing resistor and capacitor. This not only simplifies the circuit but, as will be seen, also simplifies its timing expression. In this circuit, C_t is charged through R_t and is discharged through R_1. If R_1 is made a very small fraction of R_t—that is, 1/100 or less—the output high time, t_1, will occupy almost the total timing period, T. This is because the very short period, t_2, is such a small fraction of T. Under these conditions, the timing ramp is an exponential ramp with a very short retrace. The output is a series of short negative pulses of width t_2, which occur at the basic frequency (or repetition rate).

Since t_2 is much less than T, the timing expression for t_1 can be simplified to $t_1 \cong T$. As t_1 will be $0.693R_tC_t$, this can also be the expression for T. Thus, the simplified astable expression is

$$T \cong 0.693R_tC_t.$$

Frequency is the reciprocal of T, or

$$f \cong \frac{1.44}{R_tC_t}.$$

Within these guidelines, the circuit can be operated over a fairly broad range of frequencies. For best accuracy, time t_2 should be maintained greater than 50 μs. This time is approximately 0.693 R_1C_t, and R_1 and C_t can be optimized to maintain (or exceed) this minimum pulse width.

5.11 GATED SIMPLIFIED ASTABLE

A version of the simplified astable circuit configured for positive output pulses is shown in Fig. 5-19. Here, the positions of R_1 and R_t are interchanged, with R_t governing the output low time, t_2, and R_1 governing the output high time, t_1. As in the previous circuit of Fig. 5-18, R_t is made much greater than R_1, so that the portion of the total timing interval governed by R_t will be dominant. Under these conditions, t_2 approximates T, so it may be said that $T \cong 0.693R_tC_t$.

In this version, timing period t_1 is a very small fraction of T; therefore, errors in its value will be reflected as a very small percentage of the total period, T. Because of this, the inaccuracy and temperature instability introduced by D_1 in charging C_t can be neglected for most applications, as they are scaled downward by the ratio of t_1 to T.

This circuit can also be gated by holding pin 4 low. This holds the output low, which causes C_t to discharge all the way to zero while the circuit is disabled. Normally this would cause a first cycle error as C_t charges from zero up to 2/3V+ and then subsequently charges from 1/3V+ to 2/3V+. In this case, however, t_1 is such a small fraction of the total period, T, that this error has a correspondingly small effect on the total period. The circuit may be said to be semisynchronous, since it starts up immediately with a positive-going output pulse but does have a very slight error in period T of the first cycle (compared with subsequent cycles).

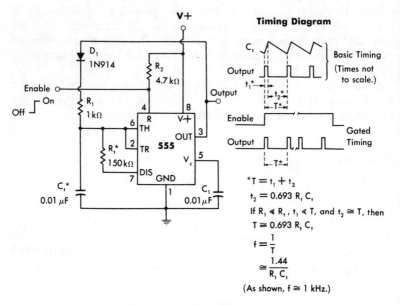

Fig. 5-19. Gated simplified 555 astable circuit.

5.12 SELECTABLE-FREQUENCY ASTABLE

For applications that require the remote programming of either of two discrete frequencies, the astable circuit of Fig. 5-20 is useful. This circuit may be recognized as a switch-selected version of the simplified astable of Fig. 5-18. Here, one of two timing resistors, R_{t_1} or R_{t_2}, is selected by the CMOS analog switch, A_2. With the control input line high, section S_1 of switch A_2 is "on," activating the R_{t_1} path to V+ and producing output frequency f_1. When the input is low, section S_2 of switch A_2 is "on," activating R_{t_2} and producing output frequency f_2.

The two output frequencies may, of course, be virtually any frequency desired within the range of the 555; here, 100 Hz and 120 Hz are shown as possible examples. This circuit produces negative-going output pulses at the pulse rate (or frequency) that is programmed. If desired, other astable circuit types can be adapted for selectable-frequency operation. For example, the gated, simplified

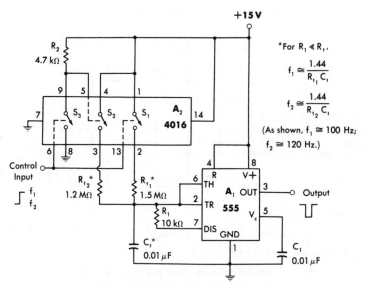

Fig. 5-20. Selectable-frequency 555 astable circuit.

astable of Fig. 5-19 can also be modified to make R_t switch selectable. Or, the square-wave astables of Figs. 5-3 through 5-7 could be modified to yield selectable-frequency, square-wave outputs.

5.13 PROGRAMMABLE-FREQUENCY ASTABLES

The concept of electronic selection of R_t can be carried a step further to yield a circuit that is programmable over a wide range of frequencies using only four switches. This circuit is also a modified, simplified astable using a 555, and is shown in Fig. 5-21. Here there are four timing resistors, R_{t_1} through R_{t_4}. It will be noted that these resistors are binary weighted; that is, each resistor is twice the value of its neighbor. This type of configuration comprises a simple 4-bit d/a converter, with the switching being accomplished by the quad CMOS analog switch, A_2. With this device, the respective switch sections close when the control inputs are high.

With four bits of control, this circuit will have $2^4 - 1$, or 15, possible output frequencies. If the basic R_t is considered the highest value (which will govern the lowest frequency), the output frequency will be

$$f_o \cong n\left(\frac{1.44}{R_tC_t}\right),$$

where n is the number from 1 to 15 as programmed by the control inputs weighted 1, 2, 4, 8, etc. In this example, R_t is 1.6 MΩ and C_t

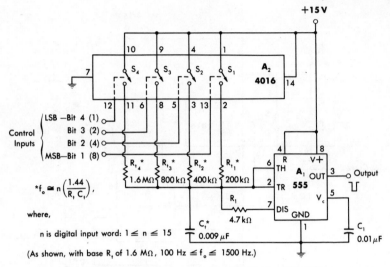

where,

$$*f_o \cong n\left(\frac{1.44}{R_1 C_1}\right),$$

where,

n is digital input word: $1 \leq n \leq 15$

(As shown, with base R_1 of 1.6 MΩ, 100 Hz $\leq f_o \leq$ 1500 Hz.)

Fig. 5-21. Programmable-frequency astable circuit using a 555.

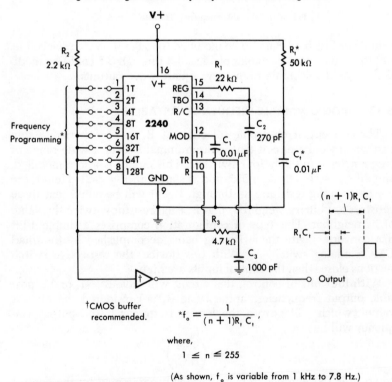

†CMOS buffer recommended.

$$*f_o = \frac{1}{(n + 1)R_1 C_1},$$

where,

$$1 \leq n \leq 255$$

(As shown, f_o is variable from 1 kHz to 7.8 Hz.)

Fig. 5-22. Programmable-frequency astable circuit using a 2240.

is 0.009 μF; therefore, f_o ranges from 100 Hz to 1500 Hz in 100-Hz steps.

Again, since this is a modified, simplified astable, the output is a negative pulse whose repetition rate is variable. This concept of wider-range programming can also be applied to other astable circuits as mentioned previously.

A programmable astable that uses the selectable-counter timing taps of the 2240 is shown in Fig. 5-22. This circuit generates a positive-going output pulse of width R_tC_t. The pulse repeats at an interval of $(n + 1)R_tC_t$, where n is the programmed number. Thus, output frequency f_o is

$$f_o = \frac{1}{(n + 1)R_tC_t}.$$

Best operation is achieved in this circuit with small C_t values such as shown; it may not work reliably with higher values. A CMOS logic buffer is recommended, used as shown. If this stage is not used, the output is inverted from that shown.

5.14 LINEAR-RAMP ASTABLES

All of the astable configurations described thus far have used an exponential timing ramp as their principle of operation. However, with the addition of a few extra parts, astables with linear time-base waveforms can be built. Linear timing ramps have many circuit uses: as time-base voltages for sweep circuits, as linear voltage-to-time converters, and many others. This section discusses methods of implementing linear-ramp astables of various circuit forms.

5.14.1 Simple Linear-Ramp Astable

One of the simplest means of producing a linear sawtooth waveform is to add a constant-current charging source to a 555 astable as in Fig. 5-23. Here, Q_1 is the added current source—a high-gain, low-leakage, pnp transistor. A suitable type for Q_1 is the 2N3906 shown or the 2N2907. The base of this transistor is referenced to the $2/3$V+ voltage at pin 5. The emitter resistance, R_t, will then set up a current in Q_1 that is proportional to V+. This current is I_t, and for a 15-volt supply, it will be approximately 4.4 V/R_t.

During time t_1, Q_1 charges C_t up to the $2/3$V+ voltage threshold of the 555, and R_1 discharges C_t down to $1/3$V+ during time t_2. The voltage difference (or ramp amplitude) is then $1/3$V+; or, for a 15-volt supply, 5 volts. The timing period, t_1, will then be

$$t_1 = \frac{5 C_t}{I_t}.$$

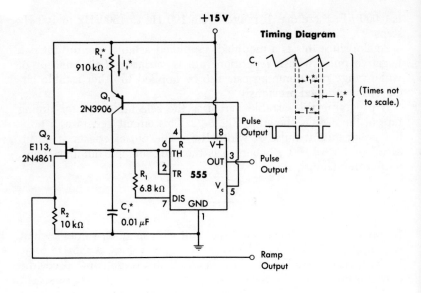

*For V+ = 15 V, $I_t \cong \dfrac{4.4\,\text{V}}{R_t}$

$$t_1 = \frac{5\,C_t}{I_t}$$

$$\cong 1.1 R_t C_t$$

$$t_2 \cong 0.7 R_1 C_t$$

If $t_2 \lll t_1$,

$$T \cong 1.1 R_t C_t$$

$$f_o = \frac{1}{T}$$

$$= \frac{0.91}{R_t C_t}$$

(As shown, $f_o \cong 100$ Hz.)

Fig. 5-23. Simple linear-ramp astable circuit.

Since

$$I_t \cong \frac{4.4\,\text{V}}{R_t},$$

$$t_1 \cong 1.1 R_t C_t.$$

The time, t_2, is set much smaller than t_1 by design and will be about $0.7 R_1 C_t$. This equation is not precise, as the current I_t is "on" during time t_2. Since t_2 is much less than t_1, t_1 is approximately equal to T, the total period. Then it can be said that

$$T \cong 1.1R_tC_t.$$

This circuit, like the basic 555 astable, is essentially supply independent. It does, however, work best at the higher supply voltages; thus, 15 volts is recommended. The design equations are approximate, due to several error sources: the V_{BE} of Q_1, the asymmetry of t_1/t_2, comparator threshold, and so on. It is a very simple circuit, however, and a good vehicle for study or experimentation. If the ramp is to be used externally, it must be buffered. A suitable buffer is an n-channel FET, such as the E113 or 2N4861 shown.

5.14.2 Improved Linear-Ramp Astable

A circuit that improves upon the simple linear-ramp astable of Section 5.14.1 is shown in Fig. 5-24. This version is similar in output, delivering positive-going output ramps and negative pulses; however, it accomplishes the task in a different manner. In this circuit C_t is charged by Q_1, an n-channel JFET current source. The constant current, I_t, produced by Q_1 charges C_t through D_1 during the output high period, t_1. When the output is low, during t_2, diode D_2 shunts current I_t from Q_1 to ground. This gating of current I_t by steering diodes D_1 and D_2 eliminates the error during timing period t_2, making t_2 dependent only on R_1 and C_t.

The voltage ramp produced across C_t is 5 volts in amplitude, as in the previous circuit, and the basic timing expression is also the same:

$$t_1 = \frac{5C_t}{I_t}.$$

However, I_t is now governed by the gate-to-source voltage of FET Q_1 and timing resistor R_t:

$$I_t = \frac{V_{GS}}{R_t},$$

for high values of R_t. For low values of R_t, I_t approaches the saturation current (I_{DSS}) of the FET. Unfortunately, V_{GS} in this expression is not readily predictable, and will vary from unit to unit due to normal FET production tolerances. Therefore, it is best to use a potentiometer for R_t in this type of circuit to compensate for V_{GS} variations. The FET used, if other than the E230, should have a gate-to-source cutoff voltage of 2 volts or less to operate properly. For design guidance, an average V_{GS} is on the order of 1 volt. (This will vary somewhat with current, of course.)

With the component values shown, t_1 is variable from less than 50 μs to over 2 ms, and t_2 is nominally 50 μs. The value specified for

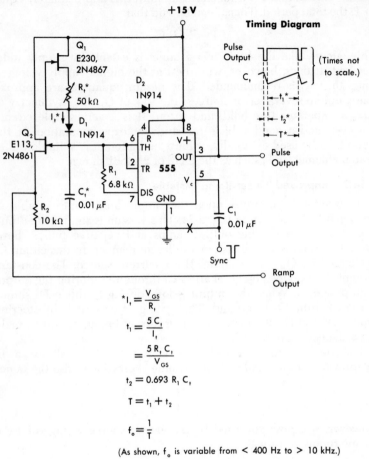

$$*I_t = \frac{V_{GS}}{R_t}$$

$$t_1 = \frac{5\,C_t}{I_t}$$

$$= \frac{5\,R_t\,C_t}{V_{GS}}$$

$$t_2 = 0.693\,R_1\,C_t$$

$$T = t_1 + t_2$$

$$f_o = \frac{1}{T}$$

(As shown, f_o is variable from < 400 Hz to > 10 kHz.)

Fig. 5-24. Improved linear-ramp astable circuit.

R_t is not an upper limit, and it may be taken higher for a greater spread in the t_1 ramp time.

If desired, the circuit can be synchronized to an external pulse source by breaking the connection of C_1 at point "X" and applying negative pulses to pin 5. Since this will terminate the positive-going ramp early, t_1 should be adjusted for a period slightly longer than the input period.

5.14.3 Negative Linear-Ramp Astable

The circuit of Fig. 5-25 is a virtual mirror image of the circuit of Fig. 5-24, delivering negative-going output ramps and positive-going output pulses. This is accomplished with virtually no change in the component count, just some rearrangement. In this circuit, C_t is charged from the output via D_1 and R_1 during period t_1. Due to the presence of D_1 and an output voltage less than V+, the t_1 timing equation is approximately $t_1 = R_1 C_t$. However, as this is both imprecise and unstable, t_1 should be minimized to reduce its overall contribution (and thus the percentage of error) to the period T. During t_2, C_t is discharged by Q_1, which is gated by pin 7 of the 555. The timing expression for t_2 is

$$t_2 = \frac{5 R_t C_t}{V_{GS}}.$$

A potentiometer is also used in this circuit for R_t to compensate for V_{GS} variations (and thus variations in I_t) of FET Q_1. With the

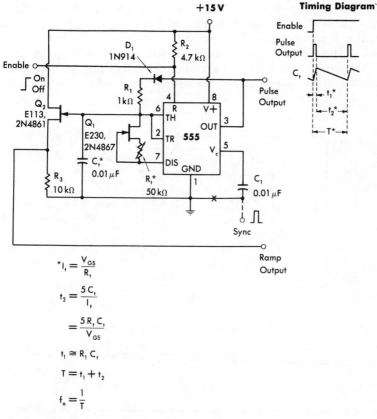

$$*I_t = \frac{V_{GS}}{R_t}$$

$$t_2 = \frac{5 C_t}{I_t}$$

$$= \frac{5 R_t C_t}{V_{GS}}$$

$$t_1 \cong R_1 C_t$$

$$T = t_1 + t_2$$

$$f_o = \frac{1}{T}$$

(As shown, $t_1 \cong 10 \ \mu s$; t_2 is variable from $50 \ \mu s$ to 2.5 ms.)

Fig. 5-25. Negative linear-ramp astable circuit.

143

component values shown, t_1 is 10 μs and t_2 ranges from about 50 μs to over 2 ms. The circuit can easily be gated "on" and "off" by the use of the reset pin as shown. Oscillations start immediately, and if t_1 is short compared to T, there will be little error caused by the gating. The circuit can also be synchronized by breaking the C_1 connection at point "X" and applying positive pulses to pin 5.

5.14.4 Wide-Range Square-Wave/Triangle-Wave Generator

The circuit of Fig. 5-26 employs linear-ramp techniques in an interesting manner, using a single constant-current source for both the charge and discharge of C_t. The result is the generation of time-symmetrical ramps, or triangle waves, as well as square waves.

In this circuit, the charge and discharge currents for C_t must come through the diode bridge formed by D_1–D_4. Bridge D_1–D_4 consists of four general-purpose switching diodes that serve to steer current in the proper direction through the current source made up of Q_1 and R_t. The unique feature of this type of current source is that it is a two-terminal device and needs no external bias connections. Thus, it serves nicely here as a floating current regulator.

The output pin serves as a source of current for the timing network, and its state of high or low determines the direction of current

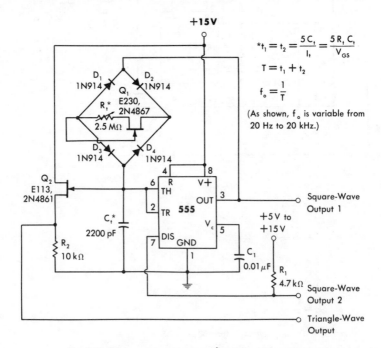

Fig. 5-26. Wide-range square-wave/triangle-wave generator.

flow into or out of C_t, for charge or discharge. Diodes D_2 and D_3 and transistor Q_1 conduct during charge while D_1, D_4, and Q_1 conduct during discharge. Since both charge and discharge currents flow through the same current regulator circuit, the currents are equal, and thus times t_1 and t_2 are equal. As a result, triangle waves are formed across C_t.

There are two square-wave outputs from the circuit, as shown. The triangle wave is 5 volts p-p in level, as set by the internal thresholds of the 555. As shown, the circuit can cover the entire audio range of 20 Hz to 20 kHz with a single 2.5-MΩ potentiometer for R_t.

REFERENCES

1. Litus, J. Jr.; Niemiec, S.; Paradise, J. *Transmission and Multiplexing of Analog or Digital Signals Utilizing the CD4016A Quad Bilateral Switch.* RCA Application Note ICAN-6601, August 1971. RCA Solid State Div., Somerville, N.J.

2. Sherwin, J. S. *The Field Effect Transistor Constant Current Source.* Siliconix Application Note, January 1971. Siliconix Inc., Santa Clara, Calif.

3. Siliconix Application Note AN73-7, December 1973. *An Introduction to FETs.* Siliconix Inc., Santa Clara, Calif.

6

IC Timer Systems Applications

This final chapter of the book is concerned with various types of applications using IC timers. In general, the chapter illustrates how IC timers can be used to satisfy particular problems, usually of a systems nature.

The general ground rules for the discussions are the same as were stated in the introduction to Chapter 4. And, as in the previous circuits, the emphasis is on high performance with maximum cost effectiveness. For those readers interested in researching the subjects of the chapter further, a list of pertinent references is included at the end of the chapter.

6.1 USING TIMERS FOR LOGIC FUNCTIONS

Both the 555 and 322/3905 timer types are highly suitable for various forms of logic functions that do not necessarily involve timing. Some of these simple but very useful applications are described in this section.

6.1.1 Schmitt Trigger

In Fig. 6-1, the 555 is shown in one of its simplest applications—as a Schmitt trigger. This circuit merely connects the two comparator inputs together and uses them as a common signal input. When the input crosses 2/3V+, the output will fall low (see the waveform diagram of Fig. 6-1). The output will then remain low until the input drops to 1/3V+, where the output will revert to a high state.

Due to the internal feedback of the 555, the output change of state is rapid and positive, and is independent of the input rate of change. Thus, output transitions will always be clean and noise-free, without "chatter" or other oscillations at the change-of-state points. The circuit will not respond to signals with p-p amplitudes of less than 1/3V+; the signal peaks must pass through 2/3V+ and 1/3V+ for operation. If desired, the output can be strobed "off" by holding the reset pin low, via the strobe input. If not used, the reset input should be connected to V+.

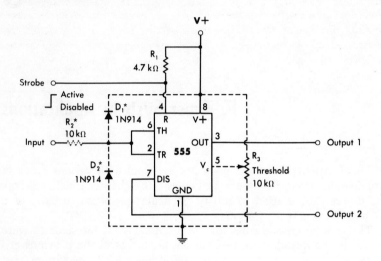

Timing Diagram

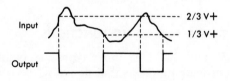

*Include R_2, D_1, and D_2
if input signal peaks are greater
than V+ or less than ground.

Fig. 6-1. Schmitt trigger.

If the input signals have peak amplitudes greater than V+ (or less than ground), protective diodes should be used with some series input resistance, as shown. The switching thresholds of the circuit can be varied if desired by adding a threshold control, as shown. With this control, the upper threshold will be the potential at pin 5, while the lower threshold will be one-half the potential at pin 5.

6.1.2 Inverting Bistable Buffer*

A modification of the Schmitt trigger is shown in Fig. 6-2. This circuit is an inverting bistable, and can also be used to amplify or regenerate signal waveforms. In this circuit, R_1 and R_2 bias the input to $\frac{1}{2}V+$, midway between comparison thresholds. Therefore, an input signal of only $1/6V+$ is required to trip either comparator. For any input above the threshold, the output is a high-level waveform at pin 3, with an uncommitted output at pin 7. Note that this circuit responds only to edges of the waveform, due to the differentiation.

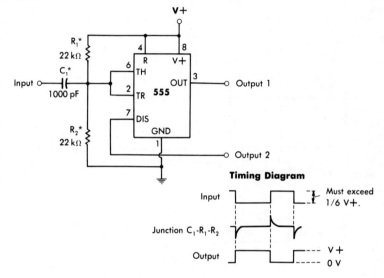

*Time constant RC $\leq 10\mu s$ (C = C_1; R = $R_1 \parallel R_2$).

Fig. 6-2. Inverting bistable buffer.

It may be noted by the reader that the 555 used alone without the differentiating network will also invert and buffer signals applied to pins 2 and 6 in common. If timing error is of no consequence, the circuit can be used simply in this fashion, but output delay may be as high as several microseconds when the input is driven to ground. The input differentiation network minimizes this delay to well below a microsecond. Component values of the differentiator are not highly critical but should be selected for an RC time constant of 10 μs or less.

---

* W. G. Jung, "Applications for the IC 'Time Machine'," *Popular Electronics*, January 1974.

6.1.3 Set-Reset (R-S) Flip-Flop

The 555 can also be used as a set-reset (R-S) flip-flop, as shown in Fig. 6-3. This circuit allows independent control of the internal latch by separating the inputs, as shown. A low pulse to the set input will force the output high. The output will then remain high until reset. There are two reset inputs: pin 4 and pin 6. Pulsing either input to the level shown will return the output to the low state (see the timing diagram in Fig. 6-3, which illustrates this). The pin 4 input is overriding and can also be used as a master inhibit, which will cause the 555 to ignore further input changes when held low.

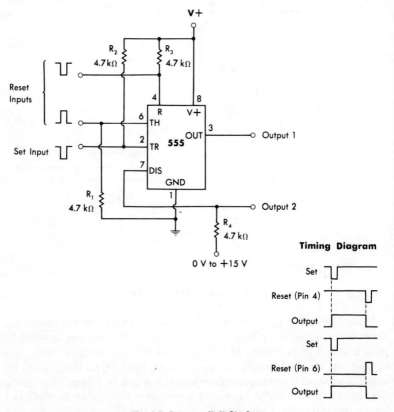

Fig. 6-3. Set-reset (R-S) flip-flop.

If the inputs at pin 6 or pin 2 are levels rather than pulses, differentiation may be required, as in the 555 monostable. Differentiation should also be used if minimum delay is required, as in the inverting bistable buffer.

6.1.4 Voltage Comparator*

The function of a voltage comparator is a very useful one, and one for which a 322 is ideally suited. Its use as such is shown in Fig. 6-4. The 322 is set up for comparator use by connecting its trigger input high, which disables the internal discharge transistor. The R/C pin is then used as a comparator input, and will compare with respect to the voltage at the V_{adj} pin (or in the case of the 3905, a fixed +2 volts). The threshold of the comparator can be varied over a range of zero to V_{ref} by means of the threshold control, R_1.

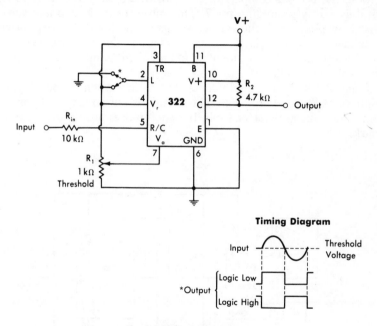

Fig. 6-4. Voltage comparator.

The sense of this comparator can be programmed by the logic pin. With the logic pin low, the output is high when the input is above threshold, and low when the input is below threshold. Grounding this pin reverses the sense. Some resistance (R_{in}) should be used in series with the comparator input if the input signal voltage has peaks greater than +5 volts or less than ground. The value used should limit the input current to 0.5 mA or less.

---

* C. Nelson, *Versatile Timer Operates From Microseconds to Hours,* National Semiconductor Application Note AN-97, December 1973. National Semiconductor Corp., Santa Clara, Calif.

Due to the high gain (and high speed with boost used), the 322 makes a very useful comparator. It can be powered from a wide range of supply voltages without threshold changes, and the output has good drive capability.

6.1.5 Zero-Crossing Detector

Fig. 6-5 illustrates a practical use of the 322 comparator—as a zero-crossing detector. Output 1 of this circuit is high when the input is above zero, and is low when the input is below zero.

Stage A_1 is a 322 comparator set up with a threshold of zero volts, established by grounding pin 7. The output of A_1 is a square wave that is in phase with the zero crossings of the input. Some positive feedback is applied via R_2 to avoid "chatter" due to noise near the zero crossings. With a value of 22 kΩ for R_1, the input can be up to ±10 volts in amplitude.

Stage A_2 is a monostable connected to fire when output 1 goes high, or at the zero crossings. This section is not an essential part of the zero-crossing detector itself, but such pulses are useful for time marks.

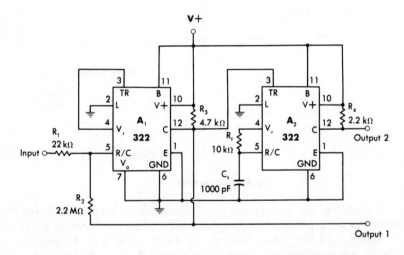

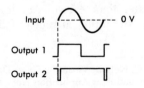

Fig. 6-5. Zero-crossing detector.

6.1.6 Window Detector

Fig. 6-6 illustrates another comparator application—a window detector. This circuit has a high output whenever the input voltage is between threshold 1 and threshold 2 in amplitude, or is within this voltage "window."

Stage A_2 is a monostable connected to fire when output 1 goes when the input is above threshold 1. Stage A_2 is an inverting comparator; its output will be low when the input is above threshold 2. With the second threshold adjusted higher than the first, the outputs can be tied together and will indicate when the input is within the window (see the timing diagram in Fig. 6-6). Both thresholds can range from zero to 3.15 volts, but threshold 2 must be higher than threshold 1 for the circuit to function properly.

6.1.7 Line Receivers

The function of a line receiver is to receive a digital signal after it has passed over a transmission line, and to regenerate it into solid

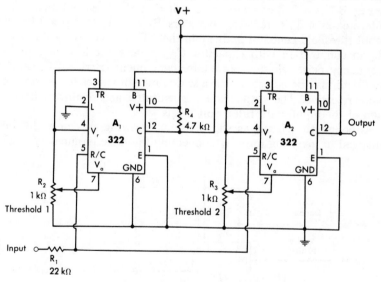

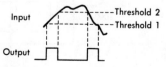

Fig. 6-6. Window detector.

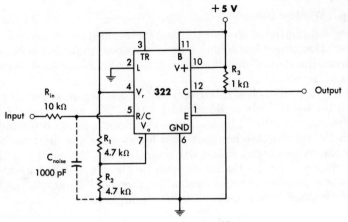

Fig. 6-7. Line receiver using 322 type.

logic levels while rejecting undesired noise components. Three circuits that accomplish this are shown in Figs. 6-7, 6-8, and 6-9.

Fig. 6-7 is a 322 comparator circuit that has been optimized for interface as a TTL receiver. Resistors R_1 and R_2 set up a 1.6-volt input threshold centered midway in the TTL output swing. Resistor R_{in} in conjunction with capacitor C_{noise} filters out incoming high-frequency noise. Neither noise-filter component is critical in value, and both may be adjusted over a wide range of values for maximum noise rejection. The output is a reconstituted and noninverted version of the input, with full 5-volt levels.

Although the circuit as shown is noninverting, it can easily be changed to inverting form by connecting the logic terminal (pin 2)

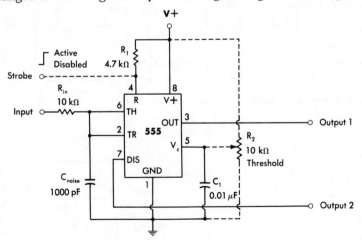

Fig. 6-8. Line receiver using 555 type.

to the reference voltage (pin 4). The circuit can also be adjusted for other logic level inputs by changing the R_1–R_2 values. Finally, the circuit can be used as a logic level converter by returning R_3 to a voltage level higher than +5 volts; for example, +15 volts for CMOS logic.

A 555 line receiver is shown in Fig. 6-8, and is similar in some regards to the 322 version. This circuit is inverting, with R_{in} and C_{noise} performing functions similar to their counterparts of Fig. 6-7. A strobe input is added here, which will hold the output low when the strobe input is low.

With a +5-volt supply, input thresholds will be 3.3 volts and 1.6 volts, respectively. This is somewhat high for a TTL logic input; thus, a threshold control may be necessary to optimize input sensitivity.

The circuit has dual outputs, and output 1 will be TTL compatible with a chip supply of +5 volts. Output 2 can be used as a logic level converter, and can be referred to voltages up to +15 volts (for any chip supply voltage).

Note that where the 322 circuit of Fig. 6-7 could convert from low to higher logic levels, the 555 circuit of Fig. 6-8 can also convert from high to lower logic levels, such as CMOS to TTL, by using a +15-volt chip supply and referring output 2 to the lower logic level.

A very useful line receiver circuit is one that responds to differential (or balanced input) drive signals. This circuit, shown in Fig. 6-9, uses both comparator inputs of the 322, with out-of-phase TTL drives applied to the (+) and (−) inputs, respectively. It can deliver an undisturbed output with about 1 volt of common-mode noise on both input lines. The output is in phase with the (+) input. As shown, the output is TTL compatible, but R_4 may be referred to higher voltages if required. Overall delay is on the order of 1 μs.

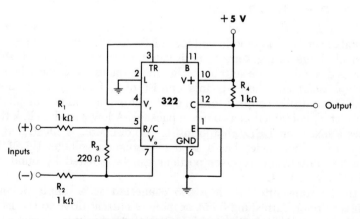

Fig. 6-9. Differential line receiver using 322 type.

6.1.8 Differential Line Driver

As a source of balanced drive signals, the circuit of Fig. 6-10 is useful. It uses the dual-unit 556, plus three discrete components. In this circuit, section A of the 556 is a noninverting buffer amplifier, while section B is an inverting bistable. The push-pull drive outputs are matched in transition times to within less than 0.5 μs.

Section B could be a simple Schmitt trigger, similar to Fig. 6-1, were it not for delay problems of the internal comparator. The differentiating network consisting of C_1, R_1, and R_2 prevents this delay by not allowing the input (pins 8 and 12) of section B to be held low. The open-collector outputs of both stages (pins 1 and 13) are also available for use, if so desired.

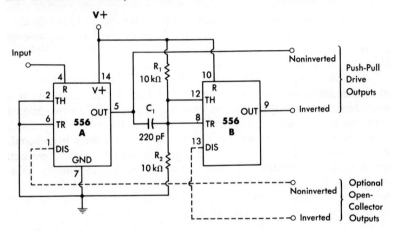

Fig. 6-10. Differential line driver.

6.1.9 Optoisolated Data Link

In general, the use of differential line transmitters/receivers is suitable for common-mode voltages of up to a few volts. When extremely large voltages are to be isolated, or complete ground isolation is desired, a more suitable technique is the use of devices known as *optoisolators*. These devices consist of a light-emitting diode (LED) optically coupled to a photodetector (photodiode or phototransistor) contained in a common package. A low-cost data link that uses a 4N37, an LED/phototransistor optoisolator, is illustrated in Fig. 6-11. This device has a 1500-volt isolation voltage; therefore, the transmitter and receiver portions can be isolated by hundreds of volts if desired.

In the transmitter, A_1 is a 555 connected as a simple Schmitt trigger driver, furnishing 5 mA or more of current drive to the LED input of the 4N37. In the receiver, the phototransistor is connected

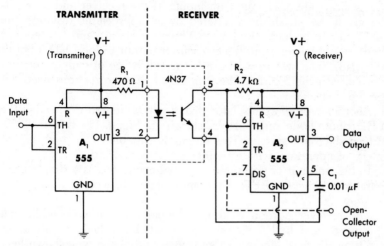

Fig. 6-11. Optoisolated data link using 555s.

as a simple common-emitter amplifier with load resistor R_2. A_2 is a second Schmitt trigger, which shapes up the relatively slow output of the phototransistor.

The system as shown delivers a high data output for a high input. Due to the slow response time of the phototransistor, delay time can be as high as 100 μs. Therefore, this circuit is suitable for rela-

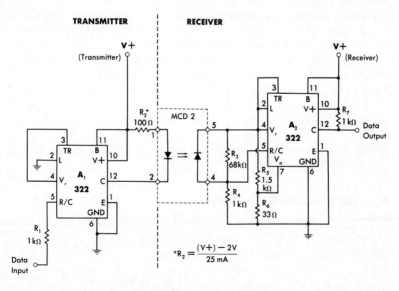

Fig. 6-12. Optoisolated data link using 322s.

tively slow rate-of-change data inputs. If the receiver and transmitter are to be physically separated by an appreciable distance, the 4N37 should be located within the receiver circuit, and twisted-pair lines should be run from the transmitter output to the LED input of the 4N37. The system works over the full range of supply voltages usable with the 555, and the supplies of the two sections need not be the same.

Another type of optoisolator that is not as limited in speed is the LED/photodiode optoisolator, shown in Fig. 6-12 within a 322-driven data-link system. This circuit is capable of a much higher speed of operation, with little increase in complexity.

In the transmitter portion, A_1 is a 322 connected as a simple comparator to drive the LED of the optoisolator. It accepts TTL inputs, as shown.

The receiver portion is also a comparator, but with additional biasing to match the photodiode output. Photodiode optoisolators are typically very low in current transfer ratio; the type shown here will typically transfer only 0.2% of the LED input current as output. Therefore, the amplifier that follows it must be capable of high gain (as well as speed). Furthermore, to realize the speed, the load resistance must be held low, which tends to minimize the useful voltage output. It is these requirements that dictate the extra complexity of the A_2 comparator stage.

Here, R_5 and R_6 set up a 70-mV bias voltage at pin 7 of the 322. With the photodiode "off," R_3 and R_4 bias pin 5 to 50 mV. A 50-mV (or more) photodiode output will overcome this bias and switch the comparator. The high gain of the 322 will then deliver a fully saturated output swing of zero volts to V+.

The system shown is noninverting overall, from data-input to data-output terminals. System delay is on the order of 2 μs, with most of this delay due to the 322 stages. The receiver can be operated from any supply voltage usable with the 322, but the transmitter should be matched to its supply voltage by selection of R_2. The R_2 value shown was chosen for a V+ of 5 volts. This system is capable of isolation comparable to that of the 4N37, as the MCD2 also features a 1500-volt isolation rating.

6.2 OUTPUT DRIVE CIRCUITS

With many timer circuits, such as the logic functions described in the previous section as well as others, it is desirable to drive higher current loads, such as LEDs, lamps, relays, etc. All of these types of loads may be driven satisfactorily using both the 555/556 and the 322/3905 types of timers. Guidelines for these modes of operation are set forth in Figs. 6-13 through 6-16.

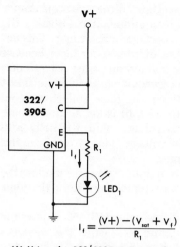

$$I_f = \frac{(V+) - (V_{sat} + V_f)}{R_1}$$

(A) Using the 322/3905 timers in the emitter-output mode of operation.

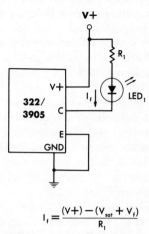

$$I_f = \frac{(V+) - (V_{sat} + V_f)}{R_1}$$

(B) Using the 322/3905 timers in the collector-output mode of operation.

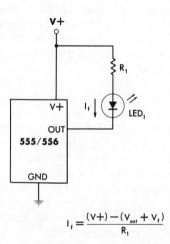

$$I_f = \frac{(V+) - (V_{sat} + V_f)}{R_1}$$

(C) Using the 555/556 timers in the LED-on-with-output-low, LED-off-with-output-high-mode of operation.

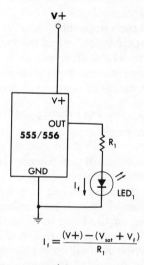

$$I_f = \frac{(V+) - (V_{sat} + V_f)}{R_1}$$

(D) Using the 555/556 timers in the LED-on-with-output-high, LED-off-with-output-low mode of operation.

Fig. 6-13. Methods for driving LEDs.

6.2.1 Driving LEDs

Fig. 6-13 shows methods for driving LEDs. In Fig. 6-13A, the 322/3905 timers are shown in their emitter output mode. Here the LED will be "on" when the output is high. Resistor R_1 is chosen for the desired LED forward current according to the equation shown. The voltage across R_1 is the supply voltage minus the 1.6-volt LED drop and the 2-volt emitter saturation voltage of the timer. This circuit is most effective at supply voltages of 10 volts or more because at lower voltages the LED and timer voltage drops are a large percentage of the supply voltage; thus the current can vary appreciably.

Fig. 6-13B shows the 322/3905 timers in their collector-output mode. The LED will be "on" when the output is low in this circuit, and the voltage across R_1 is the supply voltage minus the LED voltage drop and the collector saturation voltage. This circuit is useful over the entire range of supply voltages.

For both the emitter and collector output operating modes, the desired LED on/off state may be programmed by use of the logic pin (see Chapter 2).

In the circuit of Fig. 6-13C, which uses the 555/556 timers, the LED is "on" when the output is low, and "off" when the output is high. The voltage across R_1 is the supply voltage minus the low-state saturation voltage and the LED voltage drop. Like the circuit of Fig. 6-13B, this circuit is useful over the entire range of timer supply voltages.

In the circuit of Fig. 6-13D, which also uses the 555/556 timers, the LED is "on" when the output is high, and "off" when the output is low. The voltage across R_1 is the supply voltage minus the high-state saturation voltage and the LED voltage drop. This circuit is most useful at supply voltages above 10 volts.

6.2.2 Driving Incandescent Lamps

Although driving small panel-mounted incandescent lamps may seem to be a trivial matter, there are certain problems that can be troublesome if not completely thought out. Due to the highly nonlinear resistance/voltage characteristic of incandescent lamps, turn-on current surges can be many times the rated steady-state value for the lamp in use. If not limited, these surges can cause failure of the driver, or the lamp itself. An effective solution to this problem is to use a drive current that is peak-limited to a value somewhat higher than the normal operating current. This can easily be realized by using the 322/3905 timers with their built-in current-limiting feature, and is shown in Fig. 6-14.

In Fig. 6-14A, the 322/3905 timers are shown in the emitter output mode driving a 28-volt, 40-mA lamp. For this mode, as well as the collector output mode shown in Fig. 6-14B, the supply voltage used should be compatible with the lamp rating. In the examples shown, a 28-volt supply is used, but lower voltages (such as 12 volts or 14 volts) can also be used with lower-voltage lamps. Generally, any lamp with a 50-mA or less current rating can be used with these

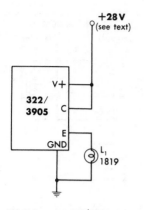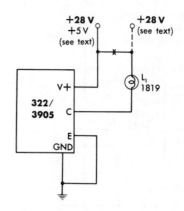

(A) Using the 322/3905 timers in the emitter-output mode of operation.

(B) Using the 322/3905 timers in the collector-output mode of operation.

Fig. 6-14. Methods for driving incandescent lamps.

two circuits. Note that in Fig. 6-14B, the V+ for the timer can be either the lamp supply or a lower-voltage supply such as +5 volts when available.

6.2.3 Driving Relays

Fig. 6-15 illustrates methods of driving relays using timers, a technique in which the total current-switching capability is extended to amperes and the total voltage-switching capability is extended to hundreds of volts. In Fig. 6-15A, the 322/3905 timers are shown in the emitter output mode. In this mode, the relay is activated when the emitter is high, and will drive the relay coil to the supply voltage minus the emitter output saturation voltage of the timer. Relays within the 322/3905 output current capability of 50 mA will generally be 24- to 28-volt types, as standard lower voltage types require appreciably higher currents. All timer-driven relay circuits should use a reverse clamping diode, such as D_1, across the coil. This diode can be a 1-A rectifier, such as one of the 1N4000 series types.

The collector output option shown in Fig. 6-15B allows a separate supply voltage to be used for the relay (if desired). In straightforward form, the relay would be returned to the timer V+, which for a 28-volt relay would require operating the timer on 28 volts also.

A useful option is to return the relay to a 28-volt supply, while the timer is operated from a lower voltage such as 5 volts. The relay supply may be any voltage up to the timer rating of 40 volts (compatible with the relay used, of course). This technique is effective in controlling noise typically generated on a relay supply line.

A 555/556 relay circuit is shown in Fig. 6-15C. This circuit will activate the relay when the timer output is low. The circuit can most

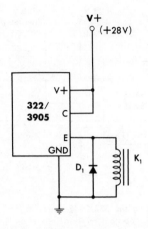

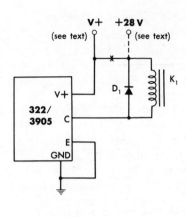

(A) Using the 322/3905 timers in the emitter-output mode of operation.

(B) Using the 322/3905 timers in the collector-output mode of operation.

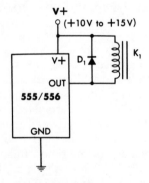

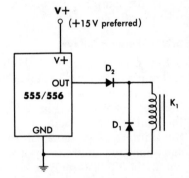

(C) Using the 555/556 timers in the relay-activated-with-output-low, relay-deactivated-with-output-high mode of operation.

(D) Using the 555/556 timers in the relay-activated-with-output-high, relay-deactivated-with-output-low mode of operation.

Fig. 6-15. Methods for driving relays.

effectively use 12-volt, 80-mA coil relays, as this combination of voltage/current matches the output drive of the 555. Supply voltage should be in the same general range; i.e., 10 to 15 volts.

The connection shown in Fig. 6-15D will activate the relay when the timer output is high, and should also be used with 12-volt relays. Due to timer output saturation drop plus the drop of D_2, this circuit should preferably use a higher voltage such as 15 volts. The purpose of diode D_2 is to prevent a timer-output latch condition in the presence of reverse spikes across the relay.

6.2.4 Booster Amplifiers

There are many situations that demand a higher current or voltage output than IC timers can handle alone, with operating speeds faster than that of a relay. This application requirement may be satisfied by the use of booster amplifier stages, as illustrated in Fig. 6-16.

In Fig. 6-16A, a 555/556 circuit for driving an npn power transistor is shown. This circuit will be "on" when the timer output is high, and the current/voltage capability is set by the device chosen for Q_1. As shown, the popular "3055" can handle up to 60 volts at currents of about 2 amperes. With this circuit, the timer V+ should be 15 volts to ensure adequate base drive for Q_1.

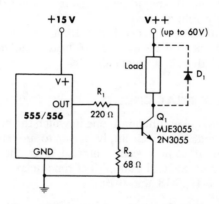

(A) Using the 555/556 timers to drive a single npn transistor.

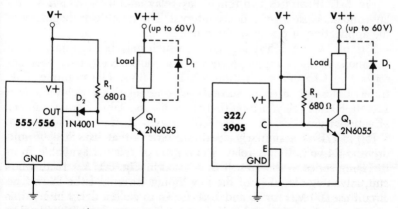

(B) Using the 555/556 timers to drive an npn Darlington power transistor.

(C) Using the 322/3905 timers to drive an npn Darlington power transistor.

Fig. 6-16. Methods for driving power booster amplifiers.

Current output capability, which is limited in Fig. 6-16A by the minimum gain of the transistor, can be increased by using a power Darlington device as shown in Fig. 6-16B. Here the 2N6055 unit can handle up to 5 amperes. Base drive is set by R_1, which should be selected in accordance with the timer V+ used for a base drive of 5 mA in Q_1. As shown, the R_1 value is for a V+ of 5 volts. This circuit relaxes the drive requirements from the timer due to the high gain of the Darlington.

A circuit arranged for the 322/3905 is shown in Fig. 6-16C. In general, this circuit is similar to that of Fig. 6-16B, but the need for diode D_2 is eliminated and the circuit has the added flexibility of on/off control through the use of the logic input of the timer. Transistor Q_1 is "on" in this circuit when the output transistor of the timer is "off." Again, R_1 should be chosen for a 5-mA drive to Q_1 with the timer supply voltage used.

In all three of the circuits in Fig. 6-16, a reverse clamping diode (shown dotted) should be used if the load is inductive in nature. This would include relays, solenoids, etc.

6.3 TIME-DELAY RELAY CIRCUITS

The time-delay relay function is one in which the closing (or opening) of relay contacts, which apply power to an external circuit, is timed to occur at some specific interval after the application of power to the timer. Several examples of time-delay relay circuits are shown in Figs. 6-17, 6-18, and 6-19.

6.3.1 Circuits for Closing Relay Contacts After a Time-Delay Period

Fig. 6-17 illustrates two time-delay relay circuits in which the relay contacts close after a delay interval. The time delay begins with the application of power to the timer.

Fig. 6-17A is a 555 delay circuit, which may be recognized as a variation of the simple power-up one-shot circuit that was described in Chapter 4 (Section 4.6, Fig. 4-6A). The example illustrated times-out after 11 seconds, closing relay K_1. Relay K_1 is a 12-volt, 80-mA relay, a type that is readily available from a number of suppliers.

Fig. 6-17B is a similarly operating circuit that uses a 3905 and drives a 24-volt, 40-mA relay. (This type of relay is available from the same series as the 12-volt units used in Fig. 6-17A.) In this circuit, advantage is taken of the low timing current of the 3905. The circuit uses 60 MΩ for R_t and 1 μF for C_t to yield a delay interval of one minute. As mentioned in Section 6.2.3 concerning 322/3905 relay drivers, the timer V+ can be lower than +24 volts if a lower voltage supply such as +5 volts or +15 volts is available.

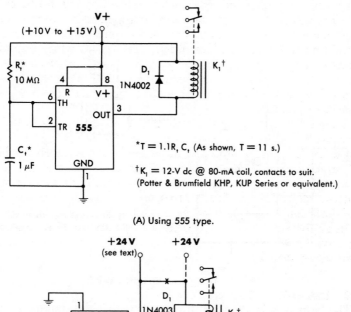

(A) Using 555 type.

(B) Using 3905 type.

Fig. 6-17. Time delay relay circuits—relay contacts close after time delay period.

Fig. 6-18[*] is an interesting variation on the relatively straight-forward circuit of Fig. 6-17B. Here, the 3905 timer is operated as a two-terminal "switch" that closes after the time-delay interval. In this circuit, the timer and its associated components to the left form the switch, while the relay to the right forms the load. Note that the collector and V+ pins of the timer are connected together, which forces the total timer current through the load. This current will be the standby current in the V+ leg, plus the collector current.

---

[*] C. Nelson, *Versatile Timer Operates From Microseconds to Hours,* National Semiconductor Application Note AN-97, December 1973. National Semiconductor Corp., Santa Clara, Calif.

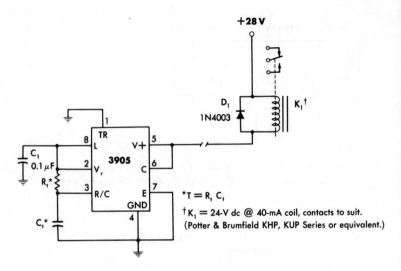

Fig. 6-18. Two-terminal time delay switch.

The basis of operation for this circuit is that the standby current of the timer prior to time-out is relatively low; in fact, it must be lower than the minimum actuation current of the relay to prevent premature closure of the relay contacts. With the application of power the timer begins timing out, and during the timing interval the timer appears as a relatively high impedance to the relay; thus, the relay remains open. When the timing interval is completed, the output stage of the timer conducts and a large current flows through the relay coil, causing the contacts to close. The relay will then remain closed until power is recycled, as in the circuit of Fig. 6-17B. This two-terminal timed switch may be used to drive loads other than relays if their current thresholds are compatible with the 322/ 3905 characteristics.

6.3.2 Circuits for Closing Relay Contacts During a Time-Delay Period

Fig. 6-19 illustrates two time-delay relay circuits in which the relay contacts are closed during a relay interval, then opened. Again, the time delay begins with the application of power to the timer.

Fig. 6-19A is a 555 delay circuit, which is also a power-up one-shot like the circuit of Fig. 6-17A. In this case, the relay is arranged to be "on" during the delay interval. This particular example is a 30-second timer and uses the same relay types as the circuit of Fig. 6-17A.

Fig. 6-19B is a 3905 "on-for-time-delay" circuit. The relay is driven from the emitter output, with the collector output connected to +24

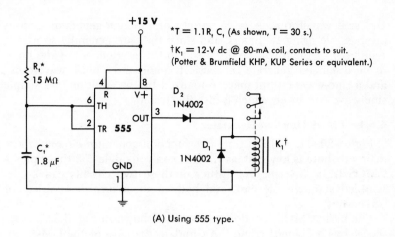

*T = 1.1R₁C₁ (As shown, T = 30 s.)

†K₁ = 12-V dc @ 80-mA coil, contacts to suit.
(Potter & Brumfield KHP, KUP Series or equivalent.)

(A) Using 555 type.

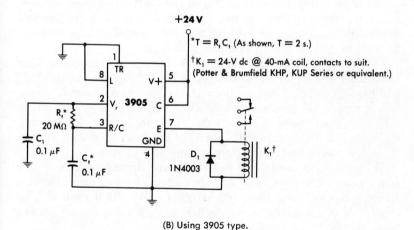

*T = R₁C₁ (As shown, T = 2 s.)

†K₁ = 24-V dc @ 40-mA coil, contacts to suit.
(Potter & Brumfield KHP, KUP Series or equivalent.)

(B) Using 3905 type.

Fig. 6-19. Time delay relay circuits—relay contacts close during time delay period.

volts. This circuit uses the same relay types as the circuit of Fig. 6-17B. As shown, it is designed for a delay of two seconds.

All of the time-delay relay circuits described in this section time-out once and only once, beginning with the application of power to the timer. As shown, they can be recycled only by the removal of power. However, if desired, the 555-based circuits can be recycled by using the reset method illustrated in Fig. 4-6A of Chapter 4.

6.4 FUNCTION GENERATORS

Some of the most useful timer circuits fall under the category of function generators, circuits that generate the basic triangle, square, and sine waveforms. A wide variety of function-generator circuits are possible using IC timers, and since they are generally so useful, a number of them will be discussed in this section. All of the types to be discussed generate the basic triangle and square waveforms, and a sine-wave output may be realized by the addition of a simple sine-wave converter, which is also illustrated.

6.4.1 CMOS Function Generator

Figs. 6-20 and 6-21 illustrate two function-generator circuits whose main attribute is low cost, as they use only a single 555 timer plus a CMOS triple inverter, A_2. Either of these two circuits can be assembled at negligible cost, and both of them furnish a variety of waveforms.

The basic CMOS function generator is shown in Fig. 6-20. It uses the 555 as a Schmitt trigger, A_1, and section A_{2B} of the CMOS inverter as an integrator. Section A_{2A} of the CMOS is a logic inverter, and these three portions of the circuit make up the basic function generator. Section A_{2C} of the CMOS is an optional inverter that can be used to provide a fourth output, if desired.

This circuit uses the CMOS and 555 characteristic supply-voltage ratiometric principles to advantage, as the voltage thresholds of A_1 and the logic level of A_{2A} are proportional to V+. As a result, operation is supply independent and the circuit operates from +5 volts to +15 volts. It can be voltage controlled, if desired, by applying a control voltage to the V_c terminal (pin 5) of the timer. If voltage control is used, it works best at a supply voltage of +15 volts.

This circuit can operate up to 100 kHz and down to extremely low frequencies since the CMOS inverters have very low input current. Resistor R_t can range from 100 kΩ up to 100 MΩ if desired. The amplitude of the triangle-wave output is $V_c/2$ volts; or, if V_c is not used, simply V+/3 volts.

Another version of this circuit is shown in Fig. 6-21, which features a linear, frequency-tuning control, R_2. In this circuit, R_1 and R_2 divide down the voltage applied to R_t, thus increasing the effective value of R_t and lowering the frequency in so doing. The frequency is lowered by the ratio of $(R_a + R_b)/R_a$, where R_a and R_b are the portions of the R_1–R_2 voltage divider as noted in the diagram. However, this type of control is effective only over about a 10:1 range, due to the offset of the CMOS inverters.

The same general statements concerning allowable values of timing components and frequency range that were made for the circuit of Fig. 6-20, also apply to this circuit. In both circuits, predictability and waveform symmetry are somewhat inexact due to variations in the CMOS threshold characteristics; therefore, the design equations are approximate.

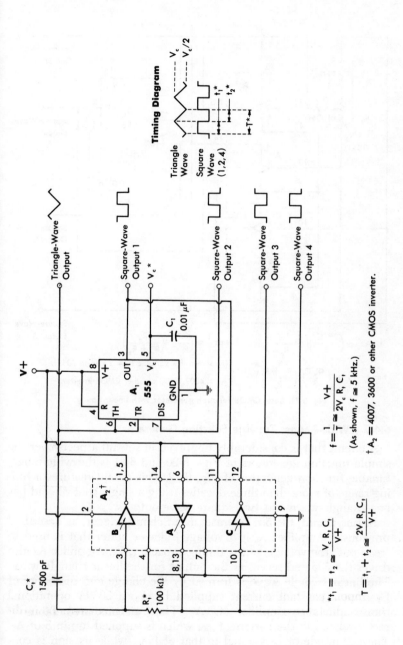

Fig. 6-20. Basic CMOS function generator with voltage-controlled period.

$$t_1 = t_2 \cong \frac{V_c R_t C_t}{V+}$$

$$T = t_1 + t_2 \cong \frac{2V_c R_t C_t}{V+}$$

$$f = \frac{1}{T} \cong \frac{V+}{2V_c R_t C_t}$$

(As shown, $f \cong 5$ kHz.)

† $A_2 = 4007$, 3600 or other CMOS inverter.

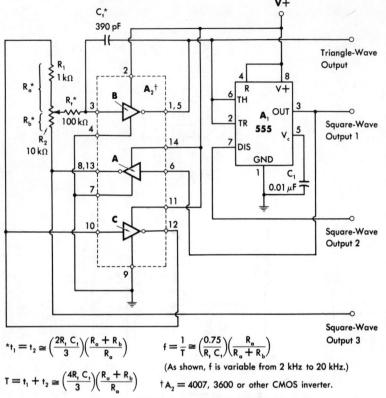

$$^*t_1 = t_2 \cong \left(\frac{2R_tC_t}{3}\right)\left(\frac{R_a + R_b}{R_a}\right)$$

$$f = \frac{1}{T} \cong \left(\frac{0.75}{R_tC_t}\right)\left(\frac{R_a}{R_a + R_b}\right)$$

$$T = t_1 + t_2 \cong \left(\frac{4R_tC_t}{3}\right)\left(\frac{R_a + R_b}{R_a}\right)$$

(As shown, f is variable from 2 kHz to 20 kHz.)

$^\dagger A_2 = 4007, 3600$ or other CMOS inverter.

Fig. 6-21. Basic CMOS function generator with linear timing.

6.4.2 Wide-Range, Tunable Function Generator

A circuit that is considerably improved in several aspects over the simple function generators of Figs. 6-20 and 6-21 is the wide-range, tunable function generator of Fig. 6-22. This circuit features a tuning range of more than three decades using a single control, and has one triangle-wave and two square-wave outputs.

In this circuit, the 555 operates as a Schmitt trigger, as described previously. Amplifier A_3 is a voltage-follower buffer that is used to "read out" the voltage across timing capacitor, C_t. Amplifier A_3 also drives the 555 and serves as the output for the linear triangle wave. The linear triangle wave is formed by the charge and discharge of C_t from a constant current supplied by A_2, a 3080A operational transconductance amplifier (OTA). The timing current, I_t, is directly related to the current, I_{ABC}, which is supplied to pin 5 of A_2. The magnitude of I_t is equal to that of I_{ABC}, while its sign is controlled by the relative state of the differential inputs (pins 2 and 3)

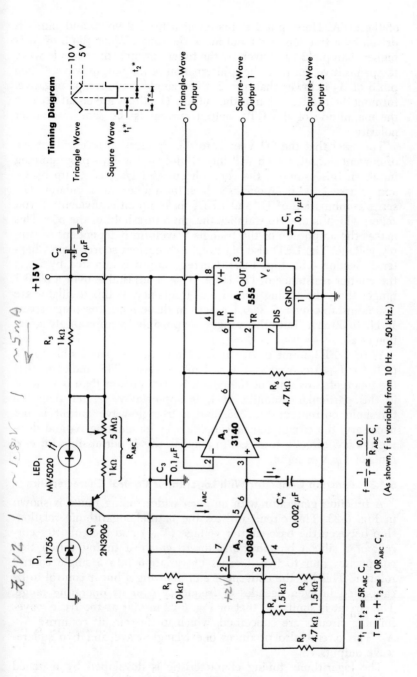

Fig. 6-22. Wide-range, tunable function generator.

171

of the OTA. Here, pin 2 is biased at a fixed 2 volts, and pin 3 is driven by a fraction of the output of the timer. When pin 3 of A_2 is higher than pin 2 (as driven by the timer output), the triangle wave ramps positive, forming the t_1 alternation of the output wave. When pin 3 of A_2 is lower than pin 2, the triangle wave ramps negative, forming the t_2 alternation of the output. The times are equal because the magnitude of the OTA output current is the same for either polarity.

The asset that the OTA lends to this application is a wide range of current output, which will linearly follow the I_{ABC} programming input. In this example, the constant current, I_{ABC}, is set up by Q_1 and its associated components, which form a current regulator. The series combination of D_1 and LED_1 make up an equivalent 10-volt zener, which is used to stabilize the pin 5 threshold of the 555. This makes the amplitude of the triangle waveform independent of supply voltage. The LED also biases Q_1 as a current source (see Chapter 4, Section 4.13, Fig. 4-20). Current output is adjusted by R_{ABC}, the emitter resistance of Q_1. Here, R_{ABC} is adjustable over a 5000:1 range; thus, the timing period and frequency of this oscillator are also adjustable over a 5000:1 range. In this circuit, the range spans 10 Hz to 50 kHz, but other ranges are possible by appropriate selection of C_t.

With a FET-input device utilized for A_3, the permissible range of I_t (or I_{ABC}) can go as low as 1 nA (or less). This indicates that the inherent capability of the circuit is even greater than is realized in this particular example. If I_t is appreciably greater than the threshold current of the 555, and a triangle-wave output is not necessary, the circuit can be simplified by eliminating A_3 and driving the 555 directly from C_t. In general, this is permissible for currents of 1 μA or more.

6.4.3 Function Generator With Logarithmic Control Characteristics

A function generator with an even wider tuning range is shown in Fig. 6-23. This circuit utilizes the natural logarithmic relationship between the base-emitter voltage (V_{BE}) and the collector current of a silicon bipolar transistor to create a tuning action that compresses four to five decades of variation into a single control rotation. This has the convenience of allowing a linear control to be calibrated in even decades of frequency over its operating range. The circuit is similar to that of Fig. 6-22 insofar as the basic wave-forming circuits are concerned, which in this circuit comprise A_1–A_3. This circuit also produces one triangle-wave and two square-wave outputs.

The logarithmic tuning characteristic is developed by matched monolithic transistors, Q_1–Q_4. This subcircuit forms a current gen-

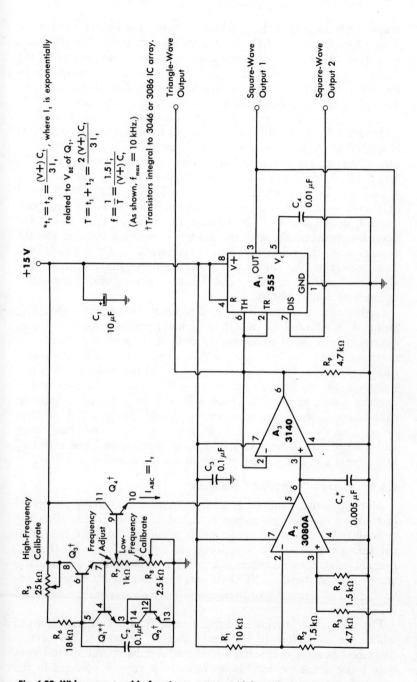

$*t_1 = t_2 = \dfrac{(V+)C_t}{31_t}$, where I_t is exponentially related to V_{BE} of Q_1.

$T = t_1 + t_2 = \dfrac{2(V+)C_t}{31_t}$,

$f = \dfrac{1}{T} = \dfrac{1.51_t}{(V+)C_t}$,

(As shown, $f_{max} = 10$ kHz.)

†Transistors integral to 3046 or 3086 IC array.

Triangle-Wave Output

Square-Wave Output 1

Square-Wave Output 2

+15 V

C_1 10 μF

A_1 555

8 V+ 3 OUT 5 V_c

4 R 6 TH 2 TR 7 DIS 1 GND

C_4 0.01 μF

R_9 4.7 kΩ

A_3 3140

C_3 0.1 μF

A_2 3080A

C_t^* 0.005 μF

High-Frequency Calibrate

Q_4† 11 9 10 $I_{ABC} = I_t$

Q_3† Frequency Adjust Low-Frequency Calibrate

R_5 25 kΩ

R_7 1 kΩ

R_8 2.5 kΩ

R_6 18 kΩ

Q_1*† C_2 0.1 μF Q_2†

R_1 10 kΩ

R_2 1.5 kΩ

R_3 4.7 kΩ

R_4 1.5 kΩ

Fig. 6-23. Wide-range, tunable function generator with logarithmic control characteristics.

erator with an output, I_{ABC}, which becomes the timing current, I_t. Transistors Q_1, Q_2, and Q_3 form a voltage regulator that produces an output of $2V_{BE}$. A reference current of nominally 500 μA is set up in Q_1 and Q_2 by resistors R_5 and R_6. This current corresponds to the full-scale I_t that yields the maximum operating frequency, and the V_{BE} of Q_1–Q_2 corresponds to this maximum frequency.

The emitter resistance of Q_3 is split into two sections: control R_7 and control R_8. A variable voltage output from frequency control R_7 is applied to Q_4, an emitter follower. Since Q_1, Q_2, and Q_4 are monolithically matched and have V_{BE}/I_C characteristics similar to the internal transistor at pin 5 of the 3080A, the collector current of Q_4 (I_{ABC}) will be exponentially related to the V_{BE} of Q_1. With R_7 at maximum resistance, I_{ABC} will be nominally equal to the reference current, or 500 μA. As R_7 is reduced, the current (I_{ABC}) in Q_4 will decrease exponentially with the linearly decreasing voltage. Viewed another way, the voltage control characteristic of Q_4 is logarithmic with respect to I_{ABC} (and I_t). The wide current range of the 3080A allows adjustment of I_{ABC} (and I_t) over a very wide range—five decades or more (100,000:1).

In this circuit, control R_8 is used to set the lower-scale operating point; it is adjusted for the desired low-frequency limit with frequency control R_7 at minimum. Control R_5 is used to calibrate the maximum operating frequency with R_7 at maximum. With the values shown, the upper operating frequency is 10 kHz, and the lower frequency limit can be, by appropriate adjustment of R_8, four or five decades lower, as desired. There will be some interaction between calibration controls, so adjustments should be repeated at least twice. This circuit is very flexible due to its wide operating range and ease of adjustment. It would make a valuable laboratory tool as a simple and inexpensive source of quality waveforms.

6.4.4 Voltage-Controlled Oscillators

A very practical function generator circuit is the voltage-controlled oscillator (VCO). This circuit is valuable for instrumentation use or in electronic music applications. It produces the basic symmetrical triangle and square waveforms as outputs, but with the frequency related to an external control voltage. There are actually many different types of VCO circuits that can use IC timers to advantage. Two examples of these circuits are shown in Figs. 6-24 and 6-25.

Fig. 6-24 is a relatively straightforward VCO circuit, variations of which are seen often in IC manufacturers' literature. The version described here is optimized for single supply operation and uses only three active devices. It operates with control voltages in the range of zero to +15 volts, and has a tuning-range capability of three

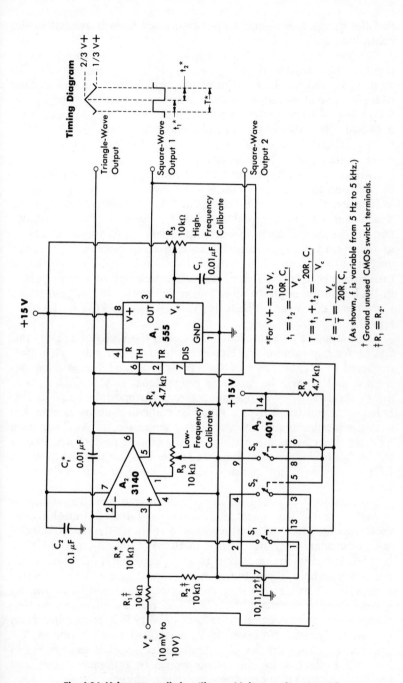

Fig. 6-24. Voltage-controlled oscillator with linear voltage control.

175

decades or more, with the output frequency linearly related to the control voltage.

The 555 timer, A_1, is used as a Schmitt trigger, which establishes input voltage thresholds of 1/3V+ and 2/3V+, respectively. Op amp A_2 is connected as an integrator, which forms a symmetrical triangle wave at its output with an amplitude of 1/3V+ (set by the 555). A_3 is a 4016 CMOS quad bilateral switch that connects timing resistor R_t either to the control voltage (V_c) or to ground. (Only three sections of the 4016 are used in this circuit; terminals for the other section should be grounded.)

In operation, with an input of V_c volts applied, a matched pair of resistors, R_1 and R_2, establish a bias voltage reference for A_2 of $V_c/2$ volts. With section S_1 of A_3 "on," R_t is connected to ground, which causes a current equal to $V_c/2R_t$ to flow in R_t. This (constant) current causes a linearly rising ramp of voltage to appear across C_t at the output of A_2. When this voltage ramp reaches 2/3V+, the output of the 555 goes low, switching section S_2 of A_3 "on," which in turn connects R_t to V_c. This again causes a current of $V_c/2R_t$ to flow in R_t, but in the opposite direction, which causes the voltage across C_t to ramp downward. When this voltage reaches 1/3V+, the cycle repeats.

The output of A_2 will oscillate between the limits of 1/3V+ and 2/3V+ for any value of V_c. However, since the current in R_t is proportional to V_c, the slope of the ramps produced (i.e., timing periods t_1 and t_2) will be *inversely* proportional to V_c, which in turn makes the frequency of operation *directly* proportional to V_c. This behavior is consistent over a range of voltages that are compatible with the input characteristics of the op amp used for A_2. Therefore, a device that can operate with a V_c near ground (a 3140) is used in this circuit to optimize the dynamic range. The fact that the 3140 is a FET-input device also permits large values of timing resistance to be used.

In order for this circuit to operate at its maximum capability, it is recommended that offset nulling, in the form of control R_3, be used with A_2. This control also serves to calibrate the low-frequency end of the operating range, at a V_c of 10 mV. Calibration of the high-frequency end of the operating range is done with R_5, which trims the 555 threshold about the nominal 10-volt point. An alternative method of achieving high-frequency calibration would be to trim R_t.

Note that the threshold levels in this circuit are referenced to V+, which will require regulation for best stability if V_c is supplied from an external source. However, if V_c is derived as a fraction of V+ (such as a voltage divider or potentiometer), variations in V+ will have little effect, since the circuit would be ratiometric. For best results, V_c should also be a low-impedance source.

A VCO function-generator circuit in which the output frequency is *exponentially* related to the control voltage is shown in Fig. 6-25. In this circuit, a *linear* increment of control voltage causes an *exponential* change in the output frequency. In practice, this type of VCO is advantageous because it can be driven with a linear time base of voltage and used with a logarithmic frequency display, as for example in frequency-response tests.

This circuit is like that of Fig. 6-23 in terms of its function-generator circuitry. Here, A_2, A_3, and A_1 are connected similarly to their counterparts of Fig. 6-23. The distinguishing feature of the circuit of Fig. 6-25 is the timing-current generator, which consists of monolithic transistors Q_1–Q_3 and op amp A_4. It is the function of this subcircuit to convert a linear voltage input into an exponentially related current, I_{ABC}, which ultimately becomes I_t. The circuit does this by again utilizing the natural logarithmic V_{BE}/I_C bipolar transistor characteristics, and also the inherent close matching between the monolithic transistors, Q_1–Q_3.

In operation, a reference current of 10 nA is supplied to transistor Q_1 through resistor R_3. With zero volts input applied to R_1, the 10 nA flowing in Q_1 will set up a V_{BE} voltage drop across it. This voltage is applied to A_4, which is a FET-input device connected as a voltage follower. The output of A_4 is connected to pin 5 of the 3080A, A_2. Thus, A_2 sees the same voltage as Q_1. Since Q_1 and the internal transistor at pin 5 of the 3080A have closely matched V_{BE}/I_C characteristics (as mentioned previously in Section 6.4.3 for the circuit of Fig. 6-23), a current identical to that in Q_1 (10 nA) will flow into pin 5 of A_2 at zero volts input.

When an input voltage is applied to R_1, the collector of Q_1 begins to rise. Because of the constant-current source driving Q_1, the voltage drop across Q_1 does not change; therefore, the incremental input-voltage changes applied to Q_1 will appear (increased by the V_{BE} of Q_1) at pin 3 of A_4, and thus also at pin 5 of A_2.

It is a physical property of matched silicon bipolar transistors that for each 10-to-1 change in collector current the base-emitter voltage must change by a factor of 60 mV (at room temperature). In the circuit of Fig. 6-25, it can be seen that an input change of 60 mV to the emitter of Q_1 will also appear as a 60-mV change to pin 5 of the 3080A, by virtue of the voltage follower, A_4. Therefore, with Q_1 operating at 10 nA, increasing the input voltage to the emitter of Q_1 by 60 mV will increase the current (I_{ABC}) to pin 5 of the 3080A by a factor of 10 (to 100 nA). Further increases in the input voltage follow the same progression; each 60-mV change in voltage to the emitter of Q_1 causes a 10-to-1 change in I_{ABC} (and thus also in I_t).

The useful range of this circuit is four decades, and it is set up for a timing-current range of 10 nA to 100 μA. With the C_t value

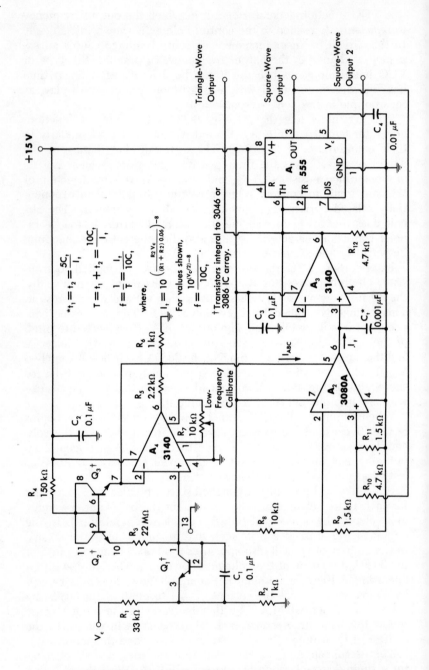

Fig. 6-25. Voltage-controlled oscillator with exponential voltage control.

shown, this yields a frequency range of 1 Hz to 10 kHz. Since an input voltage range of 60 mV per decade is a rather awkward parameter for system use, the input voltage divider, R_1–R_2, scales a larger (and more practical) input voltage range down to the required levels. As shown, the circuit has a tuning range of 2 volts per decade; i.e., zero volts = 1 Hz and 8 volts = 10 kHz. This can actually be scaled to any value desired by the appropriate choice of R_1.

Because of the nature of the current generator used in this circuit, frequency calibration of the circuit is somewhat more complex than in the previous circuit of Fig. 6-24. Offset nulling of A_4 is mandatory for calibration, due to the low voltages seen at the input. This can be accomplished initially by temporarily shorting Q_1 and applying 400 mV to pin 3 of A_4. Then, using a DVM of 1-mV sensitivity, adjust R_7 for the least voltage differential between terminals 2 and 3 of A_4. This should produce a low-end frequency close to 1 Hz. If the frequency is within 20% of 1 Hz, it can be adjusted more precisely with a final touch-up trim of R_7. Calibration of the high-end frequency is accomplished by applying an 8-volt input to R_1, and then trimming R_1 for an output frequency of 10 kHz.

6.4.5 Triangle-to-Sine-Wave Converter

The triangle-to-sine-wave converter shown in Fig. 6-26 is appropriate for use with any of the function-generator circuits described in this section. This circuit converts the constant-amplitude, triangle-wave output of the basic function generator into a low-distortion, constant-amplitude sine wave. The circuit operates on the principle of the gradual cutoff characteristics of a differential transistor pair, a concept of sine-wave shaping introduced by Grebene.[*]

In this circuit, transistors Q_1 and Q_2 are a pair of monolithically matched devices. These transistors, along with Q_3 and Q_4, are contained in a single IC array—the 3046 (or 3086). With the triangle-waveform drive to Q_1 at an optimum level, there will exist an optimum value of emitter coupling resistance, R_E, that will minimize the distortion content of the sine-wave output. Both of these conditions (i.e., optimum drive and optimum emitter resistance) must be met for minimum output distortion. With regard to the drive at the base of Q_1, the optimum triangle-wave amplitude is 346 mV p-p, and the R_1–R_2 input attenuator should be designed for this level. The R_1–R_2 values shown assume an input of 5 volts p-p; they should be altered if an input level other than 5 volts is used. With correct drive, R_E is trimmed for a distortion null in the output, and for the design shown, the R_E value will be in the range of 240 ohms.

---

[*] A. B. Grebene, "Monolithic Waveform Generation," *IEEE Spectrum*, April 1972.

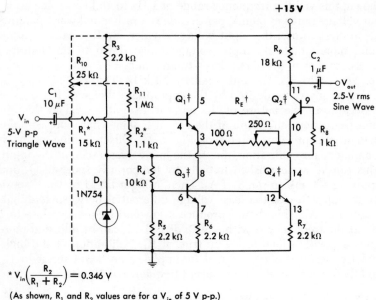

$$* V_{in}\left(\frac{R_2}{R_1 + R_2}\right) = 0.346 \text{ V}$$

(As shown, R_1 and R_2 values are for a V_{in} of 5 V p-p.)

† R_E = 240 Ω nominal. Use trim network for lowest THD (see text).

‡ Transistors integral to 3046 or 3086 IC array.

Fig. 6-26. Triangle-to-sine-wave converter.

This circuit can be used in a variety of ways, depending on the degree of performance desired. For example, a minimum complexity version could use a fixed R_E of 240 ohms, and no offset adjustment. This would result in an output THD of 1% to 2%, which is adequate for many applications. If higher performance is desired, R_E can be trimmed, as shown; also, offset control R_{10} plus resistor R_{11} can be added for output symmetry control. Adjustment of the two controls, R_E and R_{10}, will null the output distortion to 0.3% or lower, and the resulting output waveform will closely resemble a "pure" sine wave.

With the R_9 value shown, the output of this circuit is a sine wave of 2.5 volts rms. The output voltage can be lowered, if desired, by decreasing the value of R_9. The output impedance of the circuit as shown is high and is, in fact, the value used for R_9. The output impedance can be lowered, if desired, by using an emitter-follower buffer or a voltage-follower op amp connected to the output.

6.5 WIDE-RANGE PULSE GENERATOR

The circuit shown in Fig. 6-27 uses a combination of some previously described principles which, in combination, yield a high-performance, wide-range pulse generator. Both the pulse repetition rate and the pulse width are adjustable in this design over ranges exceeding four decades. The circuit consists of two functional blocks: the repetition-rate (reprate) generator consisting of timer A_1 and its associated circuitry; and the pulse generator, timer A_2 and its associated circuitry. Positive output pulses from the reprate generator trigger the pulse generator, which then produces output pulses of the desired width.

The reprate generator consists of a current-controlled 555 astable circuit, which is set up to produce positive-going, $7\text{-}\mu s$ output pulses. These pulses trigger the 322 pulse generator directly. Control of the repetition rate of these pulses controls the basic pulse-generation rate of the circuit as a whole.

In the 555 astable circuit, there are two timing capacitors, C_{t_1} and C_{t_2}. Resistor R_t and capacitor C_{t_1} set the width of the positive output pulse in the normal 555 monostable manner. Capacitor C_{t_2}, however, is charged by current source Q_4 and discharged by current source Q_3. During the $7\text{-}\mu s$ output pulse interval, current source Q_4 is gated by the diode gate, $D_1\text{–}D_2$, into C_{t_2}, charging it toward $V+$. After the $7\text{-}\mu s$ interval this current is removed, and C_{t_2} is then discharged to the 5-volt pin 2 threshold of the 555 by current source Q_3. This particular circuit arrangement causes the output pulse frequency to be a linear function of the current from Q_3. Thus, this circuit can be directly calibrated for frequency in terms of the current output of Q_3.

Transistor Q_3 is an exponential, voltage-controlled current source, similar to that previously described in Section 6.4.3, Fig. 6-23. Here, R_1 and R_2 set up the reference current in Q_1, and Q_3 conducts a percentage of this current depending on the setting of the pulse-rate control, R_3. The full-scale setting of R_3 produces the maximum frequency, and lower frequencies are produced at lower settings of R_3. The full-scale frequency is trimmed to calibration by R_1, and the low-frequency limit is calibrated by R_4, which sets the range of R_3.

The practical low-frequency limit in this circuit is about five decades below full scale when buffer transistor Q_{10} is used. If a tuning range this broad is not desired, Q_{10} and R_{16} can be eliminated, and C_{t_2} connected directly to the timer. The range of this configuration will be about three decades, and is limited by the 555. The reprate generator also has a trigger output from amplifier stage Q_5. The negative-going spikes at this output terminal can be used for external synchronization, etc.

The pulse-generator circuit also uses an exponentially variable current generator to achieve single-control, wide-range timing. In this circuit, Q_9 is the controlled current source. The pnp transistor

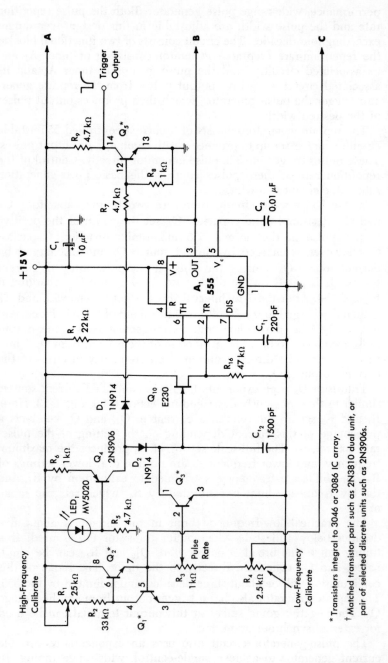

Fig. 6-27. Wide-range

* Transistors integral to 3046 or 3086 IC array.

* Matched transistor pair such as 2N3810 dual unit, or pair of selected discrete units such as 2N3906s.

† Matched transistor pair such as 2N3810 dual unit, or pair of selected discrete units such as 2N3906s.

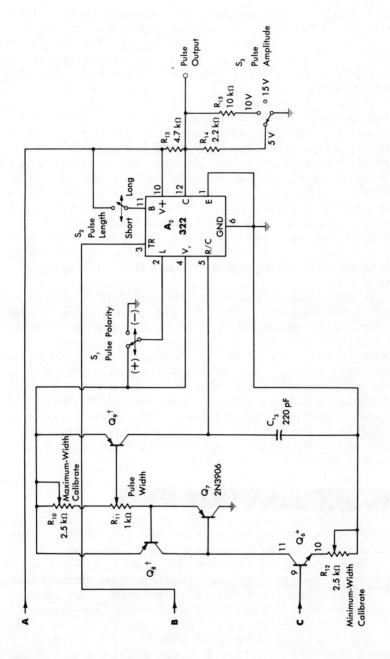

pulse generator.

pair, Q_8–Q_9, operates analogously to the npn pair, Q_1–Q_3, and produces an output current to the pulse-generator timing capacitor, C_{t_3}. The timing current is exponentially related to the V_{BE} of Q_8 in this case. This gives the pulse generator a control range of five decades or more within one rotation of potentiometer R_{11}, the pulse-width control.

The reference current for this current-generator circuit is derived from Q_1 via Q_6, a transistor that scales down the Q_1 current to the requirements of the pulse-generator circuit. The reference current flowing in Q_1 is approximately 300 μA; this current is reduced by a factor of about ten by Q_6. The final operating current for Q_8–Q_9 is set by control R_{12}, which calibrates the minimum pulse width of the circuit, with R_{11} set to minimum width. Control R_{10} calibrates the maximum pulse width, with R_{11} set to maximum. The range capability here is about five decades, if a matched transistor pair such as a 2N3810 dual unit is used for transistors Q_8 and Q_9. Alternatively, a pair of discrete transistors that have similar base-emitter voltages at 50 μA can be selected from a batch of 2N3906s or other high-gain units.

To realize best performance over the total operable range of this circuit, the boost connection of the 322 (pin 11) should be optimized for the operating pulse width. This is done via switch S_2, which should connect the boost below 1 ms and disconnect it above 1 ms. Switch S_1 programs the logic input of the 322 (pin 2) for the desired polarity of output, and switch S_3 adjusts the amplitude of the pulse output. With the values shown, the maximum operating frequency is 50 kHz, and the minimum pulse width is 10 μs. The opposite end of the respective frequency and pulse-width ranges can be set by the user as desired, and can be 0.5 Hz (or lower) for pulse rate, or 1 second (or more) for pulse width. Once set up, the operation of this circuit is stable, and there is no control interaction.

6.6 VOLTAGE-TO-FREQUENCY CONVERTERS

The voltage-to-frequency converter, or v/f converter as it is commonly referred to, has become one of the most widespread and important applications for IC timers. The v/f converter is a more developed form of voltage-controlled oscillator in which an output frequency is held precisely proportional to an input voltage. Typical accuracies are much better than 1% of full scale, with nonlinearity conversion errors of 0.05% or less often quoted. The v/f converter is a specific class of analog-to-digital converter, as it accepts an analog input voltage and produces a digital output pulse train. Dynamic ranges of representative devices are 1000:1 or more, with full-scale output frequencies of usually 10 kHz or 100 kHz.

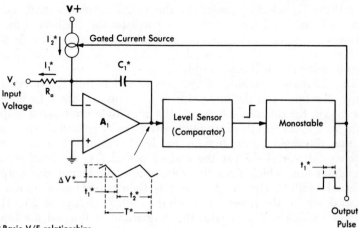

*Basic V/F relationships:

$$t_1 = \frac{\Delta VC_1}{I_2 - I_1}, \text{ and, } t_2 = \frac{\Delta VC_1}{I_1}.$$

Thus, $\Delta VC_1 = t_1(I_2 - I_1)$, and, $\Delta VC_1 = t_2 I_1$, so, $t_1(I_2 - I_1) = t_2 I_1$.

Then t_2 in terms of t_1 is:

$$t_2 = \frac{t_1(I_2 - I_1)}{I_1}.$$

Solving for total period, T:

$$T = t_1 + t_2 = t_1 + \left(\frac{t_1(I_2 - I_1)}{I_1}\right).$$

Simplifying, $T = t_1\left(\frac{I_2}{I_1}\right)$.

In terms of frequency, $f = \frac{1}{T} = \frac{I_1}{t_1 I_2}.$ In terms of V_c, $f = \frac{V_c}{R_a t_1 I_2}.$

Fig. 6-28. Block diagram illustrating basic operation of voltage-to-frequency (v/f) converter.

The basic operation of one popular type of v/f converter, known as the *charge-balancing* type, is illustrated in Fig. 6-28. The name *charge balancing* comes from the circuit operation of alternately charging a capacitor from a reference source, then discharging it with the input. The result is an output pulse train with a frequency that is linearly proportional to the input voltage.

In this circuit, C_1 is the capacitor that is charged and discharged with a constant current by virtue of the op-amp integrator, A_1. In general terms, a timing period (T) for a given frequency consists of two periods, t_1 and t_2. In this type of design, period t_1 is fixed in length and is, in fact, the pulse width of a precision monostable. Period t_2 is variable, corresponding to the input current, I_1, flowing in resistor R_a. During period t_1, the gated current source, I_2, is switched "on," and the current in C_1 will be the algebraic sum of currents I_2 and I_1. (In this circuit, current I_2 must always be greater than current I_1.) During period t_2, current source I_2 is "off," and the current in C_1 will consist only of I_1.

The alternate switching of currents I_1 and I_2 into the integrator causes the output of the integrator to ramp up and down by a voltage of ΔV, the difference between the amplitude limits. One amplitude peak of ΔV (in this case the positive peak) is fixed by the comparison threshold established by the level sensor. When the ramp reaches this threshold (at the end of period t_2), the level sensor changes state, which fires the monostable producing the output pulse of width t_1. The gated current, I_2, then causes the integrator to ramp down for the t_1 period, causing a voltage change of ΔV. This voltage change will vary with the magnitude of I_1 (and the input voltage), being maximum at lower levels of I_1.

With the application of some basic charge/voltage relationships and op-amp theory, it can be shown that the basic t_1 and t_2 relationships are:

$$t_1 = \frac{\Delta V C_1}{I_2 - I_1},$$
$$t_2 = \frac{\Delta V C_1}{I_1}.$$

Both of these equations contain the term $\Delta V C_1$ due to the charge-balancing mechanism; therefore, in terms of t_1 and t_2, the two equations can be expressed mathematically as being equal to $\Delta V C_1$ and thus equal to each other, or:

$$t_1(I_2 - I_1) = t_2 I_1.$$

Then the equation for t_2 can be expressed in terms of t_1 as

$$t_2 = \frac{t_1(I_2 - I_1)}{I_1}.$$

The equation for the total period, T, is simply the sum of t_1 and t_2, or:

$$T = t_1 + t_2$$
$$= t_1 + \left(\frac{t_1(I_2 - I_1)}{I_1}\right).$$

After some simplification, this becomes

$$T = t_1\left(\frac{I_2}{I_1}\right),$$

or, in terms of frequency,

$$f = \frac{1}{T}$$
$$= \frac{I_1}{t_1 I_2}.$$

These last two equations are the crux of the entire matter insofar as the theory of v/f converters is concerned, and warrant some discussion. For example, it can be seen from the frequency expression that frequency is directly proportional to I_1, which is the input current. Therefore, conversion sensitivity can be scaled for a given voltage by adjustment of R_a. Frequency is also inversely proportional to t_1 and I_2. Note that in terms of I_1 and I_2, the equation is ratiometric. Thus, if I_2 and I_1 are made to track one another in the same proportion, there will be no net change in frequency. This point has important practical value, as will be seen in the first actual v/f circuit described.

It should also be noted that the ΔV and C_1 terms have disappeared. The practical implication of this point is that neither C_1 nor the comparator threshold need be highly stable to maintain accuracy, as their drifts cancel during operation. The required stable elements are only those that affect I_1, I_2, or t_1. All but one of the blocks in the block diagram can be satisfied by the use of an IC timer. With the addition of a high-performance op amp, a high-quality v/f converter can be built at very reasonable cost.

Fig. 6-29 illustrates a high-performance v/f converter that accepts control voltage inputs in the range of zero to −10 volts, with −10 volts yielding a full-scale output frequency of 10 kHz. The circuit uses two 322 timers and an op amp, plus some additional components. Output 2 is a series of positive-going pulses that are nominally 15 volts in amplitude but can be programmed to other logic levels. Output 1 delivers TTL-compatible pulses of width t_1.

This circuit very closely resembles the block diagram of Fig. 6-28 in almost all respects. Amplifier A_1 is the op-amp integrator; A_2 is the comparator; and A_3 is the precision monostable. Transistor Q_1 is a switch that connects R_b to the 3.15-volt reference voltage during the t_1 timing period of timer A_3, performing the gated-current-source function. The design is based on an I_2 of 1 mA, a t_1 of 50 μs, and an I_1 of 0.5 mA for a full-scale frequency of 10 kHz.

Within the circuit, the comparator threshold of A_2 is +3.15 volts established by connecting pin 7 to pin 4, the reference output. This threshold establishes the upper limit of the ΔV output swing of A_1. As A_1 reaches the 3.15-volt threshold, the A_2 comparator output goes high, which fires monostable A_3. A_3 is a 322 timer set up for a nominal t_1 pulse width of 50 μs. During the t_1 period, Q_1 is saturated, which forces a nominal current of 1 mA through R_b. This current

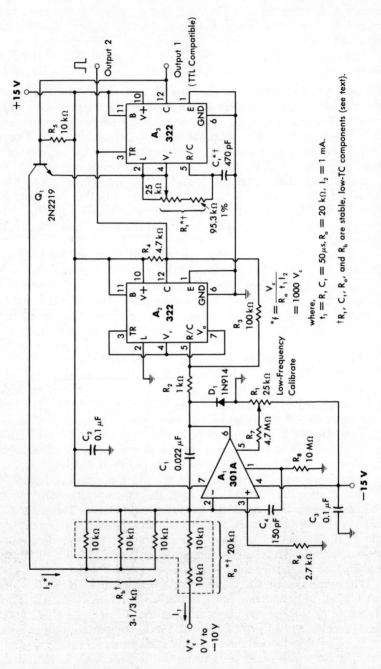

Fig. 6-29. High-performance v/f converter with negative input control voltage.

causes A_1 to ramp negative for the t_1 period. When the t_1 period has ended, I_2 is switched off and A_1 ramps positive (again) toward the 3.15-volt threshold.

There are several details of this circuit that are worthy of mention in the interest of optimizing performance. The I_1 and I_2 current-determining resistors, R_a and R_b, are best contained within a matched resistor array, preferably one with low differential matching errors and a low tracking TC. If this is done, there will be no sacrifice in accuracy and there will be a considerable savings in cost. Absolute tolerances for R_a and R_b should be 1% in order to guarantee the calibration of the circuit. As an alternative, R_a and R_b can be 20 kΩ and 3.3 kΩ low-TC discrete resistors. For best performance, R_t and C_t, which are the only other passive components that influence accuracy, should be low-TC types and of high quality in general. This would indicate a metal film resistor and a cermet trimmer for R_t, and a polystyrene or polycarbonate capacitor for C_t.

Overall calibration is achieved by nulling the offset voltage of A_1 at an input of 10 mV for low-frequency calibration, while R_t is trimmed with an input of 10 volts for a full-scale output frequency of 10 kHz. These adjustments should be repeated at least once. When calibrated, this circuit has a useful dynamic range of greater than 10,000:1, and nonlinearity is better than 0.05% of full scale.

Possible options to the circuit could include adjustments of R_a for different input voltage sensitivities, or the use of a current input connected directly to the inverting ($-$) input of A_1. Full-scale current would be 0.5 mA in this case. Due to the internal regulation, operation is essentially independent of the ±15-volt power-supply levels.

A more modest performance version of this circuit can be built by substituting a 3140 for A_1, which also allows single-supply operation (by grounding pin 4 of the 3140). The 3140 should also be nulled for low-frequency calibration.

A v/f converter that accepts positive inputs in the range of zero to +10 volts is shown in Fig. 6-30. This circuit is otherwise similar to the previous one in terms of input sensitivity, scale factor, and output pulse form. It also features a high degree of supply-voltage immunity and stable operation over a range of input voltages of three decades or more. Although this circuit does not obviously resemble the block diagram of Fig. 6-28, it does operate in a similar manner; the "blocks" of the system are just not as readily apparent.

In this circuit, C_1 is the charge-balance capacitor and is here connected to ground. The main difference in form between this circuit and the previous one is that there is no integrating op amp. Therefore, C_1 is charged and discharged by two active current sources, which produce the currents I_1 and I_2. Current I_1 is generated by A_1 and its associated components, which make up a precision voltage-

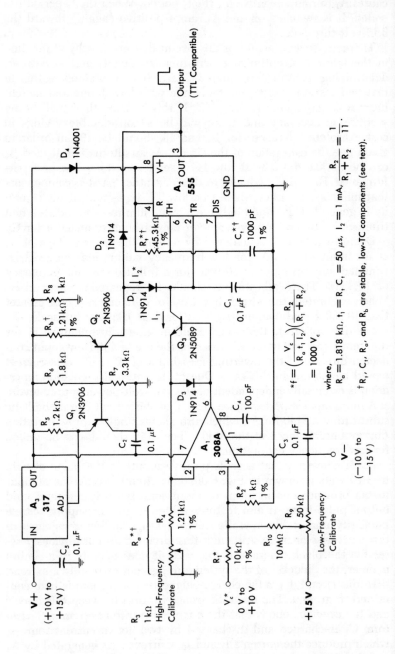

Fig. 6-30. High-performance v/f converter with positive input control voltage.

controlled current source (VCCS). The input voltage, in the 0- to +10-volt range, is divided by R_1 and R_2, and is then applied to the noninverting (+) input of A_1. The connection of Q_3 in the feedback loop forces a current through R_a that drops a voltage equal to the voltage across R_2, which is a precise fraction of the input voltage, V_c. Thus, the collector current of Q_3 is made to be a precise linear function of V_c. This collector current is current I_1, which in this case discharges C_1 during the t_2 timing interval.

In this circuit, a 555 (A_2) performs both the functions of the level sensor and the precision monostable. This is accomplished by using the 1/3V+ threshold of the pin-2 trigger input to sense the voltage on C_1, and to fire the monostable when this voltage reaches the 1/3V+ level. The pulse width produced is $1.1R_tC_t$, as in the conventional case. A further refinement is the regulation of the supply voltage to the 555, which highly stabilizes the t_1 pulse width. With the supply voltage chosen (+5 volts), this also provides an automatic TTL-compatible output.

The function of the gated current source, I_2, is accomplished by A_3, a 317 regulator, which also provides a stabilized voltage output for the 555 and A_1. The current-regulator function is provided by Q_2, which produces a nominal current output of 1 mA. This current charges C_1 during the time t_1, and is gated into C_1 by the D_1–D_2 diode gate when the 555 output is high. The voltage output of the regulator appears across R_8 and is nominally 5.6 volts, which is dropped to 5.0 volts for the 555 by diode D_4. If a TTL-compatible output is not necessary, D_4 can be eliminated (shorted) with no reduction in performance. The overall system is similar to that described in the block diagram of Fig. 6-28, except for the slope of the charge and discharge of C_1. Here, C_1 charges in a positive direction during t_1 and in a negative direction during t_2. This, however, is just a difference in application, not in basic concept.

Circuit components that influence accuracy—namely, R_a, R_b, R_t, and C_t—require good stability, as was described for their counterparts of Fig. 6-29. Also, the input divider resistors, R_1 and R_2, should be stable 1% types. The op amp used for A_1 warrants some discussion insofar as optimizing the cost/performance of the circuit is concerned. The device shown, the 308A, is a precision, low-offset, low-drift type, a consideration that is necessary because only 1 volt full scale is applied to the noninverting input of A_1. Three decades below this level, the input voltage is only 1 mV, so the offset and drift errors of A_1 can be significant if not controlled. The 308A has a typical offset of 300 μV, which allows a three-decade tuning range even without offset nulling. However, for best low-level linearity, A_1 should be nulled anyway; thus, R_9 and R_{10} are included in the circuit for this purpose. As an alternative, a 308 or a 301A can be substituted for A_1 with reasonably good performance resulting after nulling. These types will require a reduction of R_{10} by a factor of 10 in order to accommodate their higher offset voltage. Finally, if desired, this circuit can be made to operate from a single power supply by using a 3140 for A_1. This is accomplished by grounding pin 4 of the 3140 and using its regular nulling connections.

6.7 FREQUENCY-TO-VOLTAGE CONVERTERS

Closely-related to the v/f converter is the frequency-to-voltage converter, or f/v converter as it is usually referred to. This device performs a function that is just the inverse of the v/f converter: it converts an input frequency into a linearly proportional dc voltage. A block diagram of the f/v conversion process is shown in Fig. 6-31. As can be seen, many of the functional blocks are identical to those contained in a v/f converter, and, in fact, certain designs can be used for both functions.

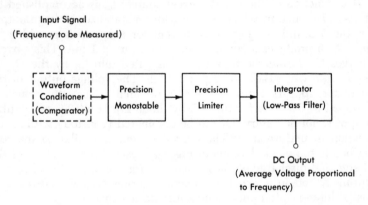

Fig. 6-31. Block diagram of frequency-to-voltage (f/v) converter.

With an input signal that occurs at some frequency, the first function shown in the block diagram is a waveform conditioner. This circuit shapes the input (which could be a sine wave, for example) into a series of sharp edges suitable for triggering the monostable shown in the next block. Since many input signals are in this form already, this block may or may not be included in a given design.

The second block is a precision monostable, which produces an output pulse of a precisely defined and stable width, regardless of the input rate. The next block is a precision limiter, which standardizes the pulse amplitude to a constant and nonvarying height. With the pulse now of standard height and constant width, the average value of its dc voltage is linearly proportional to frequency.

The final block is an integrator, which smoothes the dc time-averaged pulses into a ripple-free dc voltage. This voltage represents the input frequency, and adjustments of the monostable pulse width (or limiter amplitude) can be made to calibrate the system for specific scale factors.

Efficient design of a frequency-to-voltage conversion system is aided greatly by IC timers, which incorporate most of the capabilities outlined in Fig. 6-31. Examples of these applications are shown in Figs. 6-32 and 6-33.

In Fig. 6-32*, a 322 timer is shown in a very simple configuration that is suitable for use as a basic frequency meter, in this case a tachometer. Here, the 322 is connected as a calibrated monostable. The emitter output mode is used, with the collector tied to the reference voltage. When triggered, the output of the monostable goes high for the timing period. The pulse width is effectively regulated by the basic characteristics of the 322, while the pulse amplitude is limited to the reference voltage by the collector connection. Thus, this circuit functions as both a precision monostable and a limiter, simultaneously.

For use as a frequency meter or tachometer, a dc meter connected across R_2 will integrate the pulses, in many cases with no additional components. At low frequencies or for lightly damped meter movements, some additional filtering in the form of shunt capacitance may be necessary.

This circuit can be used as a simple tachometer to indicate engine rpm by applying the information in the table for the particular engine in use. Note that a 60-Hz reference input applied to this circuit can be used as a frequency calibration standard; for an eight-cylinder, four-cycle engine, this frequency corresponds to 900 rpm. Full-scale dc output of the circuit is 3 volts, and R_1 is adjusted to obtain a pulse width that gives the correct rpm indication. In the example shown, this would be 0.18 of full scale (for 900 Hz). Full scale in this case would be 5000 rpm.

Trigger input conditioning must be used with this circuit for tachometer use; therefore, an input network for this purpose is shown in the inset. Some adjustment of component values may be required for specific circumstances.

A circuit that is more suited to bench or lab use as a frequency meter is shown in Fig. 6-33. This circuit is identical in concept to that of Fig. 6-32 insofar as the use of the 322 is concerned. However, it also includes a post filter and scaling amplifier. Here the 322

---

* C. Nelson, *Versatile Timer Operates From Microseconds to Hours,* National Semiconductor Application Note AN-97, December 1973. National Semiconductor Corp., Santa Clara, Calif.

Basic Tachometer Information

Number of Cylinders	4	6	8
Plug Firings/Revolution	2	3	4
Plug Firings/Second @ 500 RPM	16.67	25	33.33
RPM Calibration Point for 60 Hz	1800	1200	900
Plug Firings/Second @ 5000 RPM	166.7	250	333.3

*Let $T = R_t C_t \cong \dfrac{0.95}{f_{full\ scale}}$

(As shown, $f_{full\ scale} = 333.3$ Hz; suitable for 8-cylinder engine, 5000 rpm full scale.)

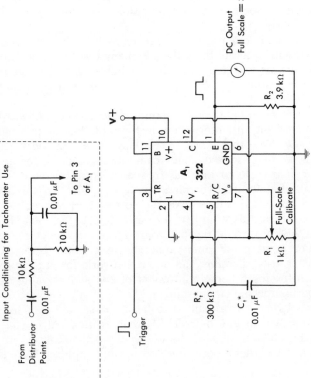

Fig. 6-32. F/v converter as basic frequency meter (tachometer).

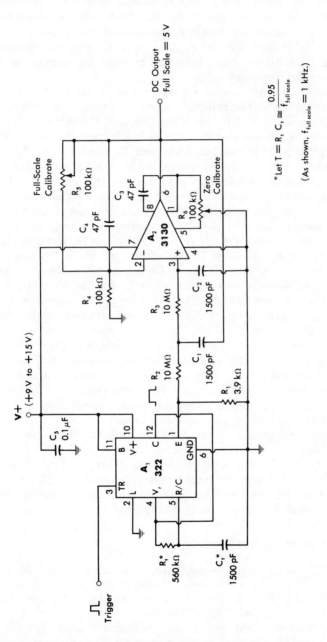

Fig. 6-33. F/v converter as lab-type frequency meter with 2-pole filter and scaling amplifier.

is wired as before, and generates positive output pulses across R_1. These pulses are integrated by the filter/amplifier consisting of R_2–C_1, R_3–C_2, and A_2. This two-pole filter effectively removes ripple due to the pulses, and the amplifier provides gain and zero adjustments. This allows exact trimming at both the upper and lower limits of the frequency range.

A 3130 op amp is used for A_2 because of its rail-to-rail output swing capability. The full-scale output is 5 volts, and with the buffering capability that the amplifier provides, this circuit can readily drive low-sensitivity meters, etc. Control R_6 provides low-frequency calibration at 1/100 of full scale, and R_5 provides high-frequency calibration at full scale. As shown, full scale is set for 1 kHz, but R_t (or C_t) can be switched for other ranges. This circuit can be battery operated, if desired, and makes a handy portable tool.

A very high accuracy f/v converter can be implemented by using the basic components of a v/f converter, as shown in Fig. 6-34. This circuit is simply a v/f converter that has been reconnected to perform the f/v function. It also follows the block diagram of Fig. 6-31, but with higher precision. The frequency to be converted is applied to monostable A_2, either directly if in the form of pulses or square waves compatible with the 322 trigger requirements, or indirectly after first being conditioned. Low-frequency or slow-rate-of-rise waveforms can be shaped by a zero-crossing detector (see Fig. 6-5), which would be the function shown in the dotted block. The standardized output pulses from the monostable drive R_b, which forms the input resistor of op amp A_1. A_1 is connected here as a scaling amplifier and filter. Resistor R_a is connected as a feedback resistor, and capacitor C_1 filters the pulse content from the output waveform, producing a smoothed dc output.

The output voltage range of this circuit is exactly the same as the input range of the v/f converter on which it is based, namely, zero to −10 volts. The full-scale frequency is 10 kHz, and all other design parameters and stability considerations are simliar to Fig. 6-29. The circuit can be optimized for dynamic range by calibrating with R_1 at the low end, and with R_5 at the high end.

6.8   DC/DC CONVERTERS

One problem that often arises in system design is the "odd" power-supply voltage. This is usually a negative voltage for one particular circuit, when 99% of the system uses a positive voltage. Another example is a voltage higher than the standard positive supply bus of the system. With just a few additional components, 555-type timers can be operated as transformerless dc-to-dc converters that can simply and easily satisfy many of these low-power, odd-voltage

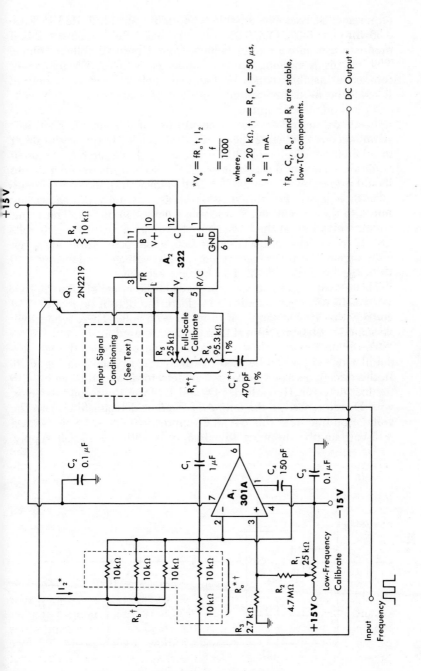

Fig. 6-34. High accuracy f/v converter.

requirements. Examples of this type of use are illustrated in Figs. 6-35 through 6-39. Fig. 6-35 is a very basic dc/dc converter that produces a negative output voltage from a positive voltage source. This circuit is nothing more complicated than a 555 minimum-component astable with a peak-to-peak detector across its output. It could be used, for example, to derive the negative supply voltage for a number of op amps.

There are several keys to successful use of this circuit. First, it is advantageous to operate the oscillator at a high frequency in order to minimize the value of filter capacitance. For example, the oscillation frequency of this circuit is 10 kHz. Second, the rectifier diodes should preferably have a low forward voltage drop in order to minimize their losses. For this reason, the germanium types shown are suited to the current levels available from a 555 output. Third, this circuit works best at the higher supply voltages; i.e., from +10 volts to +15 volts. This is because the 555 totem-pole output does not quite swing the full range of the supply voltage, and below +10 volts, the losses due to this fact become excessive.

This circuit is suitable for light loads in the range of 10 to 20 mA. With a 15-volt supply and a 1-kΩ load, the output is about −12.4 volts; with a 10-volt supply, it drops to −7.5 volts. These figures will be somewhat less if 1N914 silicon diodes are used for D_1 and D_2.

To obtain higher voltages from this type of circuit, the voltage-doubler technique can be used to advantage, as shown in Fig. 6-36. In this circuit, a second peak-to-peak detector is added in series with the first detector. Here, diodes D_3 and D_4 form the second detector, with D_3 referenced to the output of the first detector filter, C_2. The voltage output from this circuit is approximately −24 volts with a +15-volt supply, dropping to −14.5 volts with a +10-volt supply.

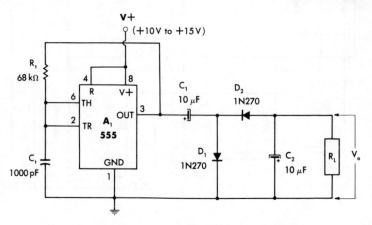

Fig. 6-35. Dc/dc converter with negative output voltage.

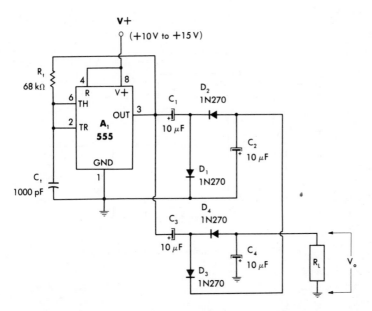

Fig. 6-36. Dc/dc converter with negative output voltage doubler.

Load current capability is approximately 10 mA, with slightly poorer regulation due to the additional series impedances. One simple way to improve the regulation is to add a low-power, three-terminal regulator such as the 79L series on the output of this circuit. Available standard voltages for the 79L series are −3, −5, −12, and −15 volts.

A circuit arranged for an output voltage more positive than the input is shown in Fig. 6-37. This circuit is like Fig. 6-35, except that the rectifier diodes are reversed and the positive rectified output voltage is "stacked" atop the input V+ line. The output voltage is about 27 volts with a V+ of 15 volts and about 17.5 volts with a V+ of 10 volts. Current capability is comparable to the circuit of Fig. 6-35—in the 10–20-mA range. An interesting option is available with this circuit by adding a low-power, three-terminal positive regulator from the 78L series. With a 15-volt regulator, this circuit can actually deliver a regulated +15-volt output from a nominal 15-volt source that is unregulated.

The two circuits shown in Figs. 6-38 and 6-39 add voltage regulation to the previously described oscillator/rectifier circuits, increasing their usefulness. In Fig. 6-38, a circuit that is useful in the range of −5 to −8 volts output is shown. This circuit uses the free collector output of the 555 (pin 7) to drive an external switching transistor, Q_1. The positive excursion of the square-wave drive to Q_1

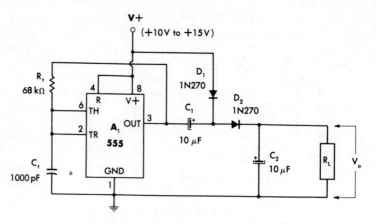

Fig. 6-37. Dc/dc converter with positive output voltage.

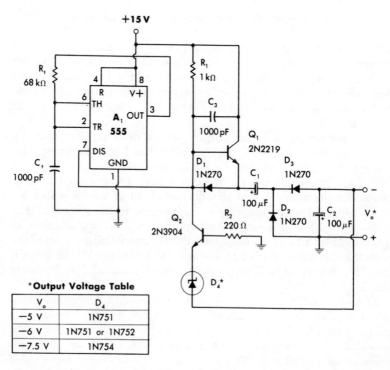

*Output Voltage Table

V_o	D_4
−5 V	1N751
−6 V	1N751 or 1N752
−7.5 V	1N754

Fig. 6-38. Dc/dc converter with regulated negative output voltages of −5 to −8 volts.

is regulated by controlling the current in Q_2. Transistor Q_2 is in turn driven by zener diode D_4, which senses the output voltage.

This scheme is quite effective in controlling the output amplitude, since the change in output voltage due to loading is considerably less than 1% for load currents of 25 mA or less. The output voltage, which is in the range of −5 to −8 volts with this circuit, is set up simply by choosing a zener diode with a voltage that, when added to the V_{BE} of Q_2, gives the desired voltage. For a −5-volt output, for example, the 1N751 is appropriate.

For output voltages greater than −8 volts, the voltage doubler of Fig. 6-39 should be used. This circuit is useful in the range of −8 to −15 volts, and as in the circuit of Fig. 6-38, the output is scaled by the choice of zener diode—in this case D_5.

6.9 FREEWHEELING, POWER-FAIL OSCILLATOR

A requirement in electronic timekeeping circuits, and other circuits that must be powered continuously, is a clock source that runs

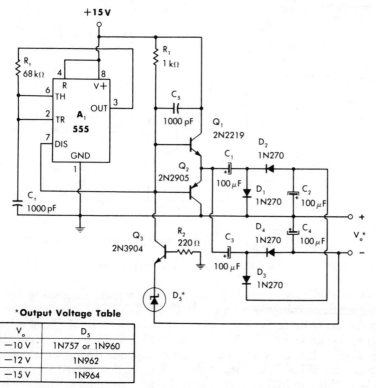

*Output Voltage Table

V_o	D_5
−10 V	1N757 or 1N960
−12 V	1N962
−15 V	1N964

Fig. 6-39. Dc/dc converter with regulated negative output voltages of −8 to −15 volts.

continuously and always at the same frequency even in the absence of primary power. Such a circuit is shown in Fig. 6-40. This circuit is either a 60-Hz amplifier or an oscillator, depending on the presence or absence of a 60-Hz ac reference signal at R_1. In normal operation, the presence of a 6.3-volt rms signal at R_1 causes an ac current to flow in the R_1–C_1 series path to ground. A portion of this input voltage appears across C_1; for the values shown, this signal will be about 2 volts p-p in amplitude.

Timing capacitor C_t would normally be returned to ground in a 555 square-wave astable. In this case, however, it is connected to C_1, which elevates the common end with the 2-volt p-p, 60-Hz reference signal. This signal passes through C_t to the timer, and is amplified by virtue of the timer acting as a Schmitt trigger. Thus, the output is a square wave that is locked to the incoming 60-Hz line frequency.

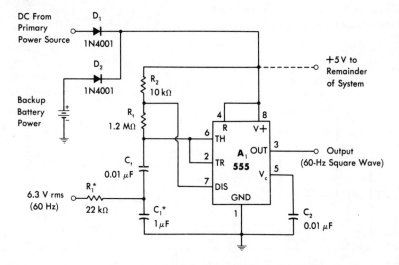

*$C_1 \gg C_t$. Choose R_t for > 1/3 V+ p-p across C_1.

Fig. 6-40. Freewheeling, power-fail oscillator.

When primary power fails, the reference voltage disappears, leaving no large voltage across C_1. With the value of C_1 set by design to be many times the value of C_t, the contribution of C_1 as a frequency-determining oscillator component is minimized, and the bottom end of C_t is now effectively at ground. Under these conditions, the 555 will oscillate at the frequency determined by R_t and C_t—in this case, 60 Hz. Timing resistor R_t should be trimmed for "zero beat" with a known 60-Hz source, prior to the connection of sync. The output for either free-running or synchronized operation is a 60-Hz square wave. The particular method of sync shown here is preferred over pulse injection at pin 5, since the R_1–C_1 network forms a natural, low-pass filter. This minimizes sensitivity to line transients, which could have disastrous effects if injected directly at pin 5. The circuit is, of course, not limited to 60-Hz applications only; it may be used at any frequency by the selection of appropriate component values. Also, the input waveform is relatively noncritical, due to the integrating effect of R_1 and C_1.

6.10   SINE-WAVE/SQUARE-WAVE OSCILLATOR

The circuit of Fig. 6-41 has the capability of producing both sine waves and square waves of low distortion from a single source. The circuit combination forms an oscillator that simultaneously generates both sine-wave and square-wave outputs, with the basic accuracy determined by the sine-wave portion. In the circuit, A_1 is a 555 timer connected in what may seem to be a minimum-component astable oscillator. This would be true if the bottom end of C_4 were grounded, and the R_6–C_4 combination would then produce a frequency of about 1 kHz. However, C_4 is returned to the output of A_2, and in operation is actually a coupling capacitor. The 555 functions

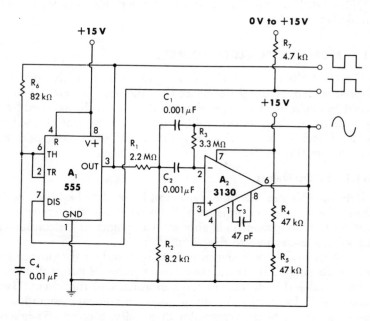

Fig. 6-41. Sine-wave/square-wave oscillator.

as an astable on start-up only; once the square-wave oscillations start, they are coupled to the A_2 circuit, which forms a multiple-feedback bandpass filter. (Design details of this type of filter appear in author Jung's *IC Op-Amp Cookbook*, published by Howard W. Sams & Co., Inc., 1974. pp. 335-336.) This filter removes the harmonic content of the square wave, producing a relatively clean sine-wave output with a distortion content of less than 2%. This sine-wave output is fed back to the timer via C_4, and the timer in this mode actually functions as a Schmitt trigger, shaping the input sine wave into an output square wave.

In addition to the sine-wave output, there are two square-wave outputs from the 555. A high-level output is available from pin 3, and an isolated, open-collector output is available from pin 7. This output may be referred to any positive voltage from 0 to 15 volts. The interesting feature of the circuit is that the center frequency of the filter is almost the sole determinant of frequency. The timer components, R_6 and C_4, have only a secondary effect, and serve primarily to start the loop oscillating at about the nominal frequency. In this example, the frequency is 1 kHz, but other frequencies can be selected by adjusting the values of C_1 and C_2 (and C_4 for large frequency changes). Resistor R_2 can be trimmed for fine-frequency changes. Resistor R_1 can be trimmed to adjust the amplitude of the output waveforms; the value shown yields an output of about 9 volts p-p.

6.11 MORE-PRECISE CLOCK SOURCES

The need often arises for a clock source that is capable of more accurate and/or stable performance than the basic RC astable timer circuit can achieve. This type of application is usually satisfied by a more stable oscillator, either an LC type or a crystal type. These two forms of oscillators can also be realized with IC timers, and are described in this section.

6.11.1 LC Oscillators

Both the 555 and 322 timer types can be used to build LC oscillators. The realized performance of the two devices is different, however, so this discussion highlights the key points to be considered in applying each timer to this application.

A 555 LC oscillator is shown in Fig. 6-42, and on the surface appears quite simple. However, there are a number of points that need to be considered in optimizing the performance of this circuit. Two basic criteria for oscillation are an in-phase feedback condition at resonance and a loop gain greater than unity. Since a 555 inverts signals between its input and its output, the resonant circuit used

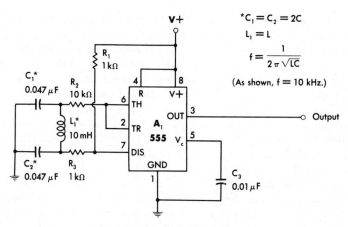

Fig. 6-42. LC oscillator using 555 type.

with it in an LC oscillator must also invert to provide an overall in-phase feedback. To provide a loop gain greater than unity, the resonant circuit cannot have excessive losses because the 555 input/output "gain" is not much more than unity (being approximately V+ divided by 1/3V+). These two criteria for an oscillator place inherent restrictions on the type of tank circuit that can be used with a 555 in an LC oscillator. Also, since whatever configuration is ultimately used should preferably be simple (as, for example, a single untapped coil), this places additional restrictions on the design.

The oscillator circuit shown in Fig. 6-42 satisfies all these objectives and uses a "pi" network as the tank circuit. This network inverts signals applied to its input from pin 7 of the 555, and the high-impedance load presented by the comparator inputs of the 555 minimizes the loading at its output, allowing high Qs to be realized. Driving this network from pin 7 of the 555 isolates the tank from the output of the 555 (pin 3), enhancing stability with respect to loading.

The circuit is set up for a given frequency by choosing L_1 and C_1–C_2 to satisfy the equation shown. There are, however, some restrictions as to the choice of component values. It will be noted that the circuit reduces to an RC astable if L_1 is shorted. The significance of this point is that the circuit does have a tendency to operate as an astable if the L/C ratio is not maintained as high as practical. Thus, in the example shown, L_1 could be 1 mH, while C_1 and C_2 could both be 0.47 μF. However, this would be a poor choice of values since the RC time constant of R_1, R_3, and C_2 would dominate and the circuit would operate as an astable. This effect can also be controlled by adjustment of R_3, which can range from 0 to 10 kΩ in value. Lower values of R_3 suppress the astable mode, while higher values (when allowable) enhance the LC filtering of the tank.

The output of the circuit is nominally a square wave, but the symmetry is not perfect due to the high voltage threshold of the 555 ($1/3V+$). The tank can develop p-p voltages in excess of V+ in operation; thus R_2 is included as an input protection resistor for the 555. The circuit can be used with any supply voltage allowable with the 555, and its operation is not strongly sensitive to supply-voltage fluctuations, although it is not completely supply independent. Due to the internal comparator delays with large input overdrive, this circuit is best used below 50 kHz.

A circuit that is highly optimized for LC oscillator use is the 322 configuration of Fig. 6-43. This circuit uses the 322 features of sensitive voltage comparison and internal regulation to advantage, and has both excellent stability and supply-voltage immunity. The 322 is connected here as an inverting comparator with the collector output loaded with a resistor to the reference voltage, rather than to V+. This regulates the amplitude of the output, making circuit operation highly immune to supply-voltage fluctuations.

The inverting comparator connection is used, which allows the previously described pi network to be used as the tank circuit. Here

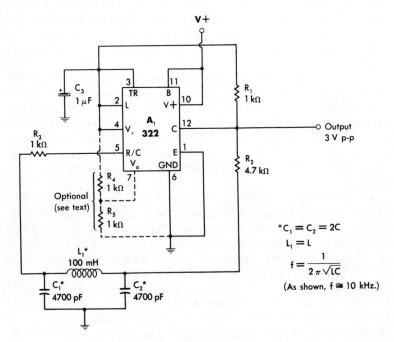

Fig. 6-43. LC oscillator using 322 type.

the comparator input is biased by dc coupling through the tank. There is essentially no shunt loading of the tank, which is an aid to achieving a high Q. The feedback signal from the tank is coupled by R_3, which prevents possible input stage damage due to high-amplitude signals (possible with some networks). With pin 7 open, the output square wave will be slightly asymmetric, because the tank dc bias of 1.6 volts is lower than the 2-volt threshold at pin 7. If this is undesirable, optional resistors R_4 and R_5 can be added. Equal values of resistance for R_4 and R_5 will yield a nominal symmetry of 50%. Or, for adjustable symmetry of the square wave, a 2-kΩ potentiometer can be used for R_4 and R_5. This option is recommended for highest stability of the circuit. The comparator, with the output clamped to the reference voltage, provides extremely hard limiting of the sine-wave input, which is an additional aid to circuit stability.

Unlike the 555 version, this circuit places no constraints on the LC ratio. Performance of this oscillator as a system will be essentially limited by the components chosen for L_1 and C_1–C_2. These components should have a high Q and a low TC (or complementing TCs) for best overall stability. The circuit works well up to about 100 kHz, where the internal delays of the 322 begin to limit the stability. Output from the circuit is +5-volt CMOS or TTL compatible; and, for best stability, the use of a buffer stage is recommended to prevent load variations from being reflected back into the circuit.

6.11.2 Crystal Oscillator/Divider Circuits

Although it is possible to configure IC timers such as the 322 directly as crystal oscillators, their performance is not optimum and is restricted to relatively low frequencies. A more efficient method of precise frequency generation is to use a CMOS inverter stage as the active portion of the oscillator, which allows excellent stability over a wide frequency range—from 10 kHz to 10 MHz. The oscillator output can be used either directly at the crystal frequency where applicable, or divided down to lower frequencies with a divider chain. There are a number of CMOS frequency dividers (counters) designed specifically for this application, and many of them conveniently include an on-chip oscillator stage. This technique is illustrated in block-diagram form in Fig. 6-44. This illustration shows a general configuration for an oscillator/divider combination. The crystal oscillator circuit will consist of an amplifier stage, plus the crystal and its associated tuning components. The oscillator output (with suitable buffering) can be used directly at the frequency of oscillation when desired. Examples of such use are microprocessor clocks, tv chrominance oscillators, and other similar time bases requiring high stability.

Since low-frequency crystals tend to be large, expensive, and not necessarily optimum in stability, an attractive and common method of generating a low-frequency clock source is a combination oscillator/divider. Such a device can use an inexpensive high-frequency crystal for the oscillator, and divide its signal down to the required final frequency with a digital divider. This technique is attractive both technically and economically, as it yields better overall stability. In addition, the resultant circuitry can be minimal in both size

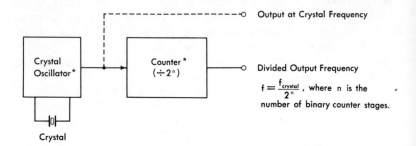

*Table of Representative Devices

Manufacturer	Part No.	Division Ratio	Comments
Intersil	ICM7038A	2^{13}, 2^{16}	Oscillator Stage
Intersil	ICM7213	2^{11}, 2^{12}, 2^{18}, 2^{22},	Programmable, With
		$2^{22} \times 60$, & mixed	Oscillator Stage
Motorola	MC14520	2×2^1, 2^2, 2^3, 2^4	
Motorola	MC14521	2^{18}–2^{24}	Oscillator Stage
Motorola	MC14536	2^0–2^{24}	Programmable
National	MM5369	2^0, & 59,659	Oscillator Stage
RCA	CD4020	2^0, & 2^4–2^{14}	
RCA	CD4024	2^0–2^7	
RCA	CD4040	2^0–2^{12}	
RCA	CD4045	2^{21}	Oscillator Stage
RCA	CD4059	3–15,999	Programmable
RCA	CD4060	2^4–2^{10}, & 2^{12}–2^{14}	Oscillator Stage
RCA	CD4017	10	Decoded 1–9 Outputs

Fig. 6-44. Block diagram of crystal oscillator/divider combination, with table of representative devices.

and cost. The table in Fig. 6-44 lists a number of devices suitable for this purpose. Virtually any required division ratio can be implemented from this table by choosing an appropriate device (or combination of devices). Many of them incorporate an on-chip oscillator stage (as noted), and some of them also feature logic-programmable division ratios. A detailed discussion of these devices is beyond the scope of the book, but a few of the more easily applied units are shown in typical circuit arrangements.

A CMOS Oscillator

A basic crystal oscillator using a CMOS inverter as the gain stage is shown in Fig. 6-45. This circuit uses a 3600 or a 4007 CMOS array with one section used as the oscillator amplifier, the other two as buffers. Section A_{1A} is connected as a linear inverting amplifier, biased by resistor R_1. This resistor, which has a value in the range of 15 to 20 MΩ, establishes a class-A bias for the stage, forcing it to operate in its linear region. The crystal, Y_1, is connected in a pi network with tuning capacitors C_2 and C_3. These capacitors are selected to yield an equivalent load capacitance, C_L, as required by the crystal in use. This capacitance is typically in the range of 12 to 30 pF. Capacitor C_3 is made variable, and trims the crystal to its specified frequency by adjusting to this capacitance.

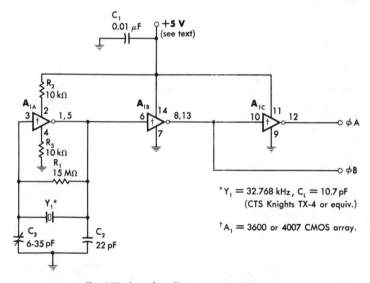

Fig. 6-45. Crystal oscillator using CMOS inverter.

This circuit will operate over a fairly broad range of supply voltages and frequencies. Best operation in terms of stability is realized when the supply voltage is constant; thus, a low-power regulator may be desirable in some cases. Although the circuit will operate at +3 volts (which would allow the use of a 322 as a regulator), this is not recommended because of parameter variations of the 3600 and 4007 units at this voltage. More reliable operation may be obtained at supply voltages of 5 to 6 volts. Within the oscillator stage itself, drain resistors R_2 and R_3 also help stabilize the oscillator against supply-voltage changes; thus, they are recommended for best performance. They can be eliminated (shorted) if the supply is regulated or if best stability is not a prime performance requirement.

Inverter sections A_{1B} and A_{1C} serve as both buffers and waveform shapers. Their output levels will, of course, be equal to the supply voltage used. Output current drive is limited for low voltage ($\cong +5$ volts) supplies to less than 1 mA. This is not sufficient to drive TTL logic without buffering, so a 4049 (or a 4050) can be used for the two buffer stages, if desired. The oscillator stage itself *must* use either a 3600 or a 4007 unit, not just any CMOS inverter. This is because some general-purpose CMOS inverters have built-in multistage buffers, and as a consequence are not suitable for linear service (4000B series units are of this nature).

If it is desired to drive logic levels higher than 5 volts, a 322 can be used as a level shifter (for frequencies below 100 kHz). The high current drive capability of the 322 allows a high fan-out, and the logic level can be referred to much higher voltages such as 10-volt or 15-volt CMOS, or even higher (up to 40 volts).

An interesting companion circuit that can be used directly with the CMOS oscillator to form an oscillator/divider combination is shown in Fig. 6-46. This circuit demonstrates the principle of binary divider devices using the already familiar 2240 units. This two-stage divider circuit uses only the counter stages of the two 2240s. The first (A_1) is a divide-by-2^8 stage; the second (A_2) is a divide-by-2^7

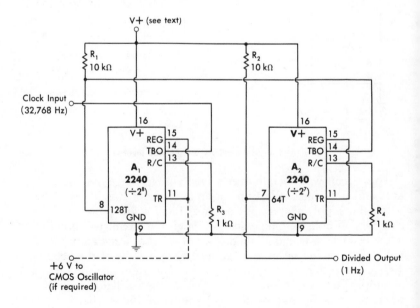

Fig. 6-46. A two-stage binary divider circuit.

stage. The composite division ratio is then their product, 2^{15}. This circuit will yield a 1-Hz square-wave output from a 32,768-Hz clock with crystal oscillator stability. This is not, of course, the limit to available frequencies, as all 16 timing taps from 2^1 through 2^{16} are available at the A_1–A_2 outputs. This factor is one major reason why a 2240 (or pair of 2240s) might be desirable for this application. A second reason is the availability of the 6-volt regulator output at pin 15, which can be used as a stabilized voltage source for the CMOS oscillator. If the oscillator circuit of Fig. 6-45 is used in this manner with a 2240, it is recommended that the 2240 supply voltage be in the range of +10 to +15 volts for best oscillator stability.

The particular design example chosen is by no means the upper limit of speed. The CMOS oscillator can operate up to 1 MHz or more, and the 2240 counters can operate at rates well above 100 kHz. This combination can therefore be operated with many varied oscillator frequencies and division ratios. It should also be pointed out that the 2250 and the 8250 counters can yield *decade* dividers, which operate similarly.

Precision Oscillators

An example of a high-precision, low-frequency clock source using the Intersil 7038A device is shown in Fig. 6-47. The 7038A functionally consists of an oscillator stage, plus a divider chain with a length of 2^{16}, with a tap at 2^{13}. There are two outputs from the final (2^{16})

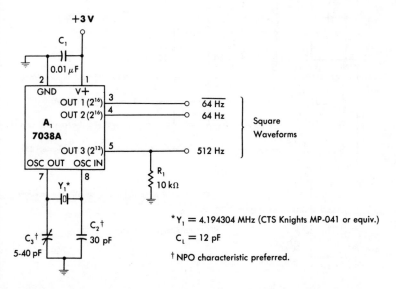

Fig. 6-47. A high-precision crystal oscillator/divider circuit using the Intersil 7038A.

divider stage in push-pull format (pins 3 and 4). The 2^{13} stage (pin 5) is a single-ended source and must use an external pull-down resistor (R_1). The drive from output pins 3 and 4 is TTL compatible (for one load or less), and pin 5 is 3-volt CMOS compatible. The device can use crystals in the range of 0.5 to 10 MHz, and operates at a +3-volt supply level. Current drain is quite low, allowing efficient two-cell battery operation; or it may use a 3-volt regulator (such as the 322).

As shown, a 4.194304-MHz crystal is used, which yields complementary 64-Hz drives and a 512-Hz drive. The crystal specified is designed for a C_L of 12 pF, which is provided by the C_2–C_3 trim network. These capacitors should be low-TC units, preferably NPO types. Overall performance of this circuit is quite high, featuring less than 0.1 ppm/°C stability in the 0–70°C range, and less than 0.1 ppm/V of supply sensitivity.

Another high-precision clock source is shown in Fig. 6-48. This circuit uses the Intersil 7213, which has four outputs, as shown. They are TTL compatible with the addition of pull-up resistors. The composite output (output 2) consists of 16-Hz and 1024-Hz waveforms, gated at a 2-Hz rate. The outputs as shown, with the crystal specified, are directly usable in electronic clocks. This de-

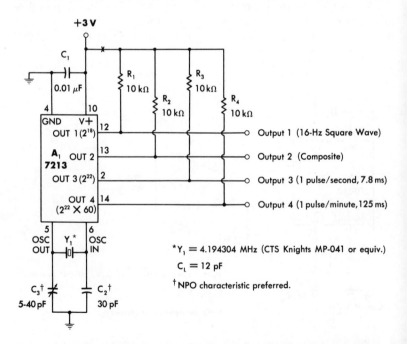

Fig. 6-48. A high-precision crystal oscillator/divider circuit using the Intersil 7213.

vice also uses a 4.194304-MHz crystal and a 3-volt supply, and overall stability is comparable with the circuit of Fig. 6-47. Pull-up resistors must be used on the active outputs, which can be referred to voltages other than the 3-volt supply. By breaking the connection at point "X" and referring R_1–R_4 to +5 volts, for example, an optimum TTL interface is obtained. The outputs should see voltages less than +6 volts in all cases.

A third high-precision clock source is shown in Fig. 6-49. This circuit uses the National Semiconductor 5369 device, which here employs a standard 3.58-MHz color-tv crystal as the oscillator, and divides the signal down to 60 Hz (pin 1). A buffered oscillator-frequency output is also included (pin 7), which can be used either for monitoring purposes or to drive a second divider chain. These outputs are CMOS compatible and swing between the supply-voltage levels.

The 5369 oscillator can be employed over a very wide frequency range, from 10 kHz to 4.5 MHz. Above a few megahertz, however, it must use supply voltages of 10 volts or more, as in the case shown here. The supply-voltage range in general is +3 to +15 volts, and a local regulator can be used for optimum stability. The oscillator bias resistor must be added externally across the crystal as shown.

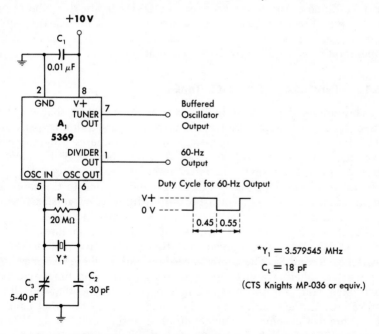

Fig. 6-49. A high-precision crystal oscillator/divider circuit using the National Semiconductor 5369.

The use of any of the devices described in this section can satisfy virtually any specific oscillator frequency requirement, either directly or by the additional use of one of the divider units. In applying these units, however, the reader is cautioned to always ensure input/output compatibility when coupling between differing devices, particularly when operating at differing supply voltages.

6.11.3 "Never-Fail" 1-Hz Clock

A basic timing requirement is a 1-Hz (or 1-pps) clock source. There are a variety of ways that such a source can be implemented, but one popular method is to derive the timing reference from the 60-Hz power line. In Fig. 6-50, a circuit is shown that produces output pulses at 1-second intervals with the basic accuracy referred to the power line. Stage A_1 is a freewheeling 60-Hz oscillator synchronized to the 60-Hz reference. The 60-Hz output of this stage is divided by A_2, an 8260 operating as a divide-by-60 counter. The 60-Hz input is fed to pin 14, with the time base disabled by connecting pin 13 to ground through R_3. The carry-out terminal, pin 15, is used as the divide-by-60 output and produces a TTL (or 5-volt CMOS) compatible logic swing.

With backup power applied to the D_1–D_2 OR gate, this circuit will continue to run in the absence of a 60-Hz input; thus, the term *never fail*. The circuit can operate over a +5-volt to +15-volt supply range, but power drain is minimized at +5 volts, making the operation more suitable for battery use at this level.

6.12 UNIVERSAL APPLIANCE TIMER

The circuit of Fig. 6-51 is a universal timer circuit that can be used to control ac-powered appliances for preset timing periods. This circuit is basically a two-stage, seconds/minutes programmed counter driven by a 1-second clock derived from the 60-Hz power line. Stage A_1 is a divide-by-60 counter, which is triggered by a 60-Hz signal developed across diode D_2 and applied to the counter input of A_1 by diode D_1. The 1-Hz output from A_1 triggers stage A_2, which in turn triggers stage A_3. Stages A_1–A_3 are all 8260s.

Stage A_2 counts seconds, while stage A_3 counts minutes. These two stages are programmed by their respective thumbwheel switches, S_1 and S_2. The common terminals of all these switches are connected together and to the load resistor, R_4. The timer will be activated (on) for the programmed period of time that this line stays low. In the standby state, the counters are in the reset condition, due to the common connection of all reset inputs to R_4. When the Start Time switch, S_3, is activated, the R_4 bus goes low and the timing cycle begins. While the R_4 bus is low during the timing

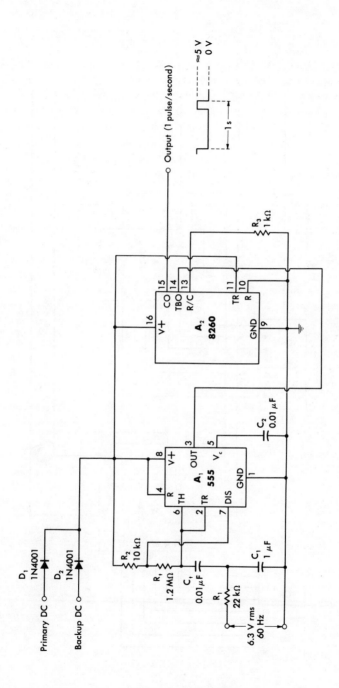

Fig. 6-50. "Never fail" 1-Hz clock.

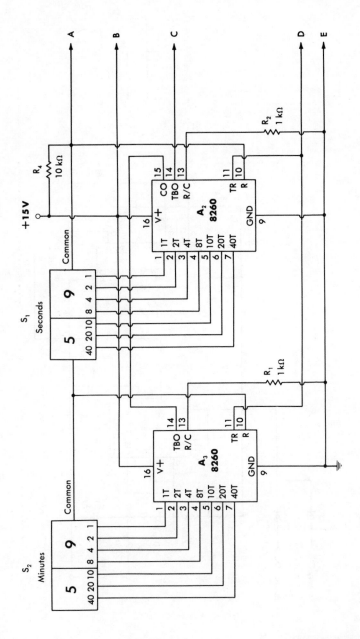

Fig. 6-51. Universal

216

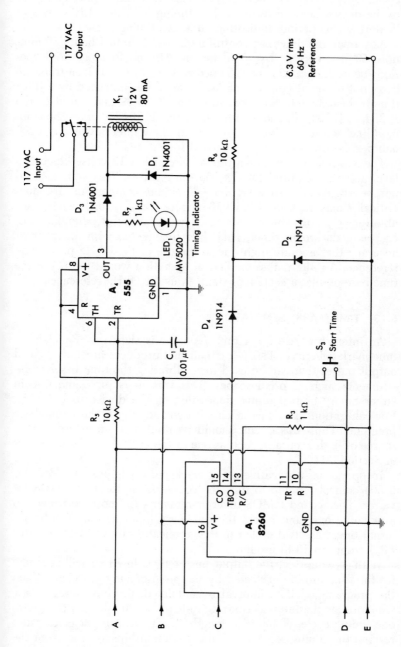

appliance timer.

217

cycle, relay driver A_4 holds relay K_1 closed, which applies ac power to the device being controlled. The timing indicator, LED_1, is also lit during this period, indicating an active timing cycle.

As shown, the timer can control intervals of up to 1 hour (59 minutes, 59 seconds) in 1-second increments. Should longer time periods be desired, a 2250 or 8250 bcd counter can be driven from the carry out terminal (pin 15) of A_3. It would be controlled by a third thumbwheel switch pair, variable from 00 to 99 and wired to the 1, 2, 4, 8, 10, 20, 40, and 80 outputs of the 2250/8250, with the common line tied to R_4. This option would allow intervals of up to 100 hours to be programmed in 1-second increments.

The supply voltage shown for the circuit is +15 volts, which is a level actually required for the 555 relay driver. This voltage need not be regulated, however, and all circuit components are noncritical. Relay K_1 should have 10-ampere contacts, or heavier, suitable for ac loads. This circuit has a wide variety of potential uses; e.g., as a darkroom timer, a kitchen timer, etc., with an inherent accuracy that is quite high due to the use of the 60-Hz power-line reference. An alternative method of operation would be to use the time-base-oscillator section of stage A_1 as the 60-Hz reference.

6.13 TIME-MARK GENERATOR

An interesting use for a timer/counter is shown in Fig. 6-52—a time-mark generator. This is a circuit that produces precisely spaced output pulses that can be used for calibrating the time bases of instruments such as oscilloscopes. This version is programmable in either binary or bcd terms, depending on the device used for A_1. For calibration uses, such a circuit demands the highest time-base precision. Thus, this circuit should be used with an external clock source of high accuracy, such as one of the crystal oscillator sources of Section 6.11.

In operation, a negative-going clock pulse triggers the 2240 (or 2250) control flip-flop to start the counters, via the first CMOS inverter. The second CMOS inverter reinverts the positive trigger to a negative one, and clocks the time-base input at pin 14. The internal time base is disabled in this application by the 1-kΩ resistor (R_2) from pin 13 to ground.

With A_1 counting, the output bus at R_1 is low and will stay low for the time duration selected by the programming switches. When the programmed time interval is completed, the bus voltage rises, which resets the internal control flip-flop via pin 10. This completes one timing cycle, which is of length T_0. The next input clock pulse restarts the timing sequence. Timing relationships are shown in the diagram. Note that the final output is inverted from the program-

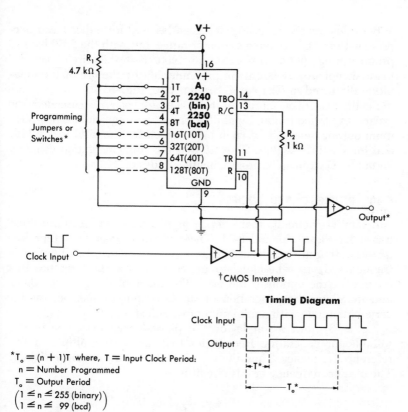

$$*T_o = (n + 1)T \text{ where, } T = \text{Input Clock Period:}$$
$n = \text{Number Programmed}$
$T_o = \text{Output Period}$
$\left(\begin{array}{l} 1 \leq n \leq 255 \text{ (binary)} \\ 1 \leq n \leq 99 \text{ (bcd)} \end{array} \right)$
(As shown, $T_o = 4$.)

Fig. 6-52. Time-mark generator.

ming bus via the third CMOS inverter, in order to reestablish logic levels.

The negative-going output pulses are of a single clock period of width T in this circuit, and will occur at time intervals of:

$$T_o = (n + 1)T,$$

where n is the number programmed into the timer/counter. This number is from 1 to 255 for the 2240, and from 1 to 99 for the 2250. Thus, T_o is variable from 2T to 256T with a 2240, and from 2T to 100T with a 2250.

This circuit is also a programmable frequency divider, as it divides the input frequency inherently. For an input frequency of f, the output frequency will be f_o, which is expressed as

$$f_o = \frac{f}{n + 1}.$$

Both devices shown in Fig. 6-52 can be used for either basic purpose, with the 2240 having a greater range, but with the 2250 having programming that is easier to use. Note that this circuit can be made completely electronic in programming by the use of the technique discussed in Chapter 4, Section 4.10, Fig. 4-14.

Finally, two (or more) of these circuits can be cascaded for greatly expanded timing capability. For a two-stage circuit, the minimum output time will be $(n + 1)^2 T$, which for the 2240 is 65,536T, and for the 2250 is 10,000T. Note that this implementation can also be made electronically programmable, if desired.

6.14 PHASE-LOCKED LOOPS

A family of circuits that operate as phase-locked loops are illustrated by Figs. 6-53 and 6-54. These circuits generate output frequencies that are referenced to an input frequency, f_r. The output frequency, f_o, may be equal to, above, or below the reference frequency, at the option of the user. The inherent flexibility of these circuits allows the generation of stable output frequencies over a very wide range with a minimum of complexity.

Fig. 6-53 shows a simple form of phase-locked loop, one that is suitable for generating frequencies that are a submultiple of the reference frequency. Thus, this circuit is termed a *frequency divider*. The output frequency of this circuit is

$$f_o = \frac{f_r}{M},$$

where M is the division ratio of the circuit and is an integer. The circuit accepts an input reference frequency in the form of a square wave, and limits its amplitude to V+ volts p-p in the first CMOS inverter, A_{3A}. The output of A_{3A} is integrated by the RC network, R_9–C_2, forming a 2-volt p-p triangle reference waveform across C_2. This triangle wave is sampled by cascaded sample/hold switches, S_1 and S_2, which are portions of a 4016 CMOS analog switch. Switch S_1 is "on" during the high output state of A_1, the local (f_o) clock oscillator. When S_1 is "on," C_{H1}, the first sample/hold capacitor, is effectively in parallel with C_2; thus, the reference triangle is impressed across C_{H1}. When S_1 switches "off," C_{H1} retains the last instantaneous voltage level of the reference waveform. Switch S_2, which is driven from the complemented output of A_1 via inverter A_{3B}, is "on" when S_1 is "off," and vice versa. During the S_2 "on" period, S_2 transfers the sampled charge on C_{H1} to C_{H2}. When S_1 again turns "on," S_2 is "off." Because of this "bucket-brigade" process of the cascaded sample/hold components—sample, hold/transfer, and sample again—C_{H2} is never allowed to see transient or large varia-

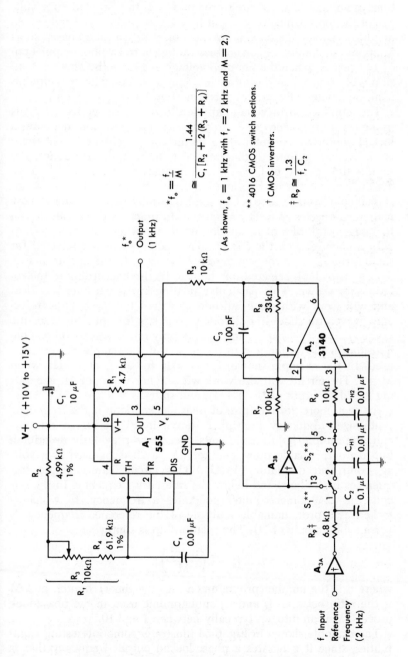

Fig. 6-53. A phase-locked loop, frequency-divider circuit.

$*f_o = \dfrac{f_r}{M}$

$\cong \dfrac{1.44}{C_t\,[R_2 + 2\,(R_3 + R_4)]}$

(As shown, $f_o = 1$ kHz with $f_r = 2$ kHz and $M = 2$.)

** 4016 CMOS switch sections.

† CMOS inverters.

‡ $R_9 \cong \dfrac{1.3}{f_r\,C_2}$

tions in voltage. The voltage stored on C_{H2} is the sampled error voltage of the loop, which is read out by FET amplifier A_2. Amplifier A_2 amplifies the voltage on C_{H2} by a factor of 4/3, a ratio chosen so as to apply the nominal +10-volt bias to the control voltage input (pin 5) of timer A_1 when the sampled voltage is +7.5 volts. This relationship will occur when the reference ramp is sampled at its midpoint, assuring a nominally correct bias condition for frequency lock.

The amplified error voltage applied to A_1 via R_5 induces changes in the center frequency of A_1 in a direction to cause and maintain a locked condition. An interesting feature of the circuit is that there is no overriding limit to the ratio of f_o to f_r, and, in practice, the sample time period of f_o can be many times a reference cycle of f_r. This is because the usable sampled error voltage is only the *last* instantaneous voltage across C_{H1}, and it is relatively immaterial how many prior cycles of f_r have occurred. In practice, this means f_o can be integer multiples of f_r, ranging up to about ten.

In a given design, the 555 oscillator components are selected for the desired f_o, as in the standard astable case. In the circuit of Fig. 6-53, resistor R_9 is selected for the f_r to be used according to the relationship shown. This particular relationship is not overly critical and will work with some latitude of values for R_9, but the reference waveform should be 2 volts p-p or less for optimum dynamic range. The circuit has an inherent supply-rejection capability, due both to the 555 design and to the nature of the sample/hold system used. In the example shown, $f_o = 1$ kHz for an f_r of 2 kHz when M = 2. The circuit will also work with higher values of M, generating a 1-kHz output of $f_r = 2$ kHz, and so on.

An even more flexible form of phase-locked loop is the frequency synthesizer/frequency multiplier shown in Fig. 6-54. In this circuit, the sample/hold components operate as previously described, but a programmable timer/counter (A_1, a 2240) is used as a voltage-controlled oscillator (VCO). This circuit can generate frequencies both above and below the reference frequency. In the circuit, the 2240 timer/counter generates a frequency, f_o, which is locked as a programmable multiple of the reference frequency, f_r (or a subharmonic of f_r). The relationship is simply

$$f_o = f_r \left(\frac{n+1}{M} \right)$$

where n is the number programmed into the timer/counter, and M is the ratio between f_r and its subharmonic used in the phase-lock process. M is an integer, typically between 1 and 10.

The example shown in Fig. 6-54 illustrates some interesting capabilities, since it generates a phase-locked output frequency that is *not* a direct multiple of the reference frequency. For $f_o = 50$ Hz,

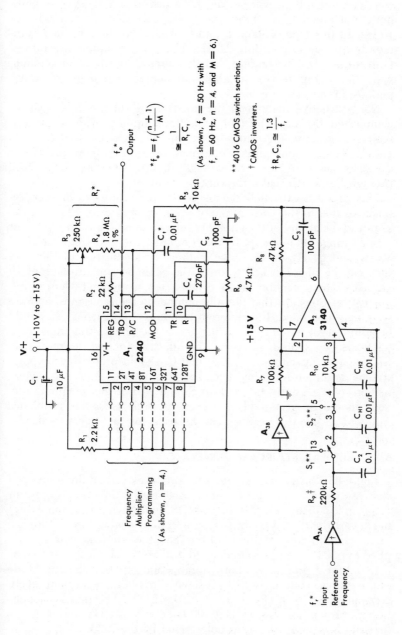

Fig. 6-54. A phase-locked loop, frequency-synthesizer/frequency-multiplier circuit.

223

n = 4, and M = 6, S_1 will be "on" for a period of $1/f_o$, every n clock periods. Here, $1/f_o$ is 20 ms, and the sample rate is $f_o/(n+1)$, or 10 Hz. In this type of loop, f_o and f_r are locked through their common frequency submultiple, 10 Hz. Thus, f_r is sampled at a rate of f_r/M, or 10 Hz. The feature that this circuit does have in common with that of Fig. 6-53 is its ability to sample over many reference periods; in this case it is 6.

Nondestructive readout of the stored error voltage on C_{H2} is accomplished by A_2, a FET-input buffer amplifier. This stage also scales the error voltage from its nominal dc base line of +7.5 volts (at C_{H2}) up to the required dc level at the control input (pin 12) of the 2240. The amplified error voltage from A_2 is applied to the 2240 through R_5, which scales the control sensitivity.

This circuit is extremely flexible in its ability to lock different frequencies due to the wide range of programming available with the 2240 and the ability of the circuit to select a sampling ratio (M) over a wide range of integers. In this example, the two frequencies of 50 Hz and 60 Hz do not have a direct harmonic relationship, yet they can be phase-locked through the common subharmonic of 10 Hz. Advantage can be taken of this fact to generate locked-frequency relationships that literally number in the thousands, and are seemingly unrelated. For different reference frequencies, R_9 should be adjusted for a 2-volt p-p swing across C_2. The 2240 nominal center frequency should, of course, be adjusted for the desired f_o via R_t (or C_t). Capacitors C_{H1}, C_{H2}, and C_t should preferably be polystyrene or other low-dielectric-absorption types. Any two switch sections of a 4016 can be used for S_1 and S_2, but the remaining sections should be grounded to prevent possible cross-talk problems. Any CMOS inverter sections can be used for the inverters shown.

6.15 BIPOLAR STAIRCASE GENERATOR

Fig. 6-55 illustrates a circuit that produces a digitally generated staircase waveform. It uses a single 2240 as the time-base generator (A_1) and a 7530 (A_2, a 10-bit, CMOS, multiplying d/a converter) to convert the digital count from the 2240 to analog form. With the bipolar multiplying capability of the 7530, the reference voltage (applied to pin 15) can be varied up to ±10 volts. As a result, the staircase output it produces is both variable and bipolar. The amplitude of the staircase output is ±10 volts, with 255 staircase steps or levels, corresponding to the 8-bit count of the 2240. The staircase repetition rate is variable from about 400 Hz down to 4 Hz, and staircase linearity is within +0.2%, as determined by the 7530.

In the circuit, the 2240 is connected as a free-running astable, with its clock frequency variable via R_t. Operation as shown here is self-

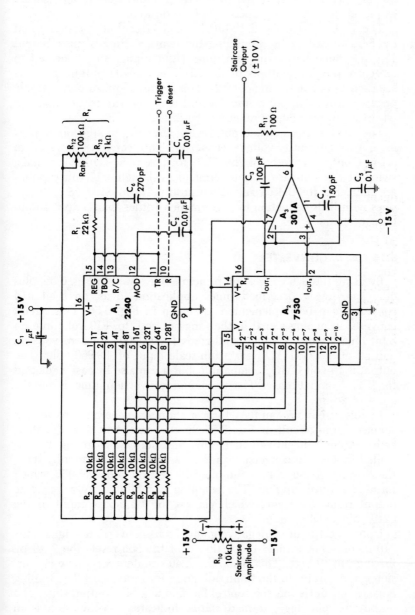

Fig. 6-55. Bipolar staircase generator.

225

starting by virtue of the connection of the trigger input (pin 11) to pin 15, but the separate use of the trigger and reset inputs can be employed for start/stop operation, if so desired.

The 7530 is basically a 10-bit d/a converter, but it can be used for 8-bit operation by grounding the two LSB inputs, pins 12 and 13, as is done here. The remaining eight inputs are connected to the corresponding outputs of the 2240, and 10-kΩ pull-up resistors are used to define the logic levels. This device requires only a single positive power supply at pin 14, and it is also directly compatible with the supply-voltage range of the 2240.

The 7530 is a current output device; thus, A_3, a 301A op amp, is used as a current-to-voltage converter. This converts the zero to ±1-mA output current of the 7530 to a zero to ±10-volt voltage swing at the staircase output terminal. Although ±15-volt supplies are indicated for the circuit, it will also perform with ±5-volt supplies with little reduction in performance (other than output amplitude, of course).

6.16 A/D CONVERTER

By using the d/a conversion principle illustrated in Fig. 6-55, plus other previously described timer application ideas, an 8-bit a/d converter can be implemented as shown in Fig. 6-56. This circuit converts an analog signal in the range of zero to +10 volts into an 8-bit binary word. Coding is all zeros (00000000) for zero volts in, and all ones (11111111) for a full-scale input of +9.960 volts. The output is 15-volt CMOS compatible, but it can also be easily adapted for TTL compatibility. The maximum conversion time is on the order of 26 ms.

In this circuit, A_2 and A_3 form a staircase generator, somewhat similar to the circuit of Fig. 6-55. In this case, however, the staircase output is negative-going only, and start/stop timing is controlled by the a/d conversion process. In the static or standby state, the counter of the 2240 is inhibited by section B of the 556, which forms a control flip-flop. The counting is inhibited by a low-state output from this stage, which clamps the time-base input of the 2240 (pin 14) to ground.

Upon receipt of a negative-going "start-conversion" signal, the 110-μs one-shot formed by section A of the 556 resets the 2240 by driving pin 10 high. This clears the 2240 counters for a new conversion cycle. (Refer to the timing diagram for sequence.) The trailing edge of this pulse sets the control flip-flop to a high output state. The output of this stage, termed *status*, indicates a conversion is in process when it is in the high state (see the timing diagram). The existence of this high state simultaneously triggers the 2240 (pin 11)

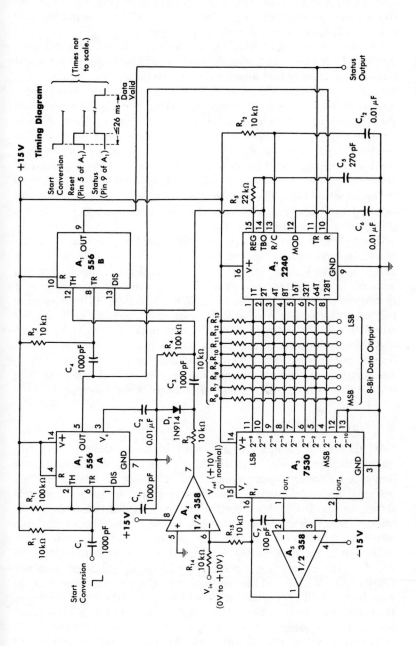

Fig. 6-56. A/d converter.

and removes the clamp from the counter input (pin 14), enabling the counters to start counting. The counters will now count upward toward full scale during the conversion cycle.

The advancing count from the counter drives the 7530 d/a converter, which produces an increasingly negative output. This negative output is compared against the input voltage, V_{in}, by comparator A_4, one-half of a 358 op amp. The reference for this comparator is zero volts, established by the ground on the (+) input. When the d/a output voltage is equal to the input voltage, the (−) comparator input crosses zero, which forces the output positive. The positive-going comparator output resets the control flip-flop through C_3, which brings the status line low, indicating a complete conversion. The 2240 output data are now valid, as the d/a output is equal to the input voltage. The negative-going edge of the status line can be used to strobe the 8-bit output data into a storage register for subsequent processing.

The system can be calibrated by either of two methods. If a stable variable voltage in the range of +9 to +11 volts is available, this can be used as V_{ref} and applied to the 7530 reference input (pin 15). This could be a divided-down, +15-volt power-supply voltage if it is sufficiently stable. Alternatively, a fixed, stable, +10-volt source can be connected to V_{ref}, and R_{15} trimmed ±10% to calibrate the output for all "1s" with +9.960 volts applied to V_{in}.

Although the level of the binary output data as shown is 15 volts due to pull-up resistors R_6–R_{13} being returned to +15 volts, these resistors (alone) can be returned to +5 volts for TTL compatibility, if desired (with no other circuit alterations being necessary).

6.17 SPEED ALARM

IC timers can accurately, and with high sensitivity, detect frequency differences within an incoming pulse train. Within calibrated limits, a frequency detector based on IC timers can be effectively operated as an over-speed or under-speed alarm. Such a circuit is shown in Fig. 6-57. This circuit uses a pair of 322 timers as a calibrated monostable and comparator, respectively. The monostable stage, A_1, produces a fixed-width, positive output pulse which appears across R_2. Resistor R_2 is returned to the regulator output of A_2; thus, the voltage drop across R_2 will be stable. Since both the supply voltage used and the width of the pulse are fixed, the average voltage that will appear across R_2 will vary linearly with frequency. It is this feature that allows the circuit to be used as a frequency-to-voltage converter.

The pulse train at R_2 is integrated by R_3–C_1 and applied to the comparator input of A_2. Since the voltage threshold of a 322 is a

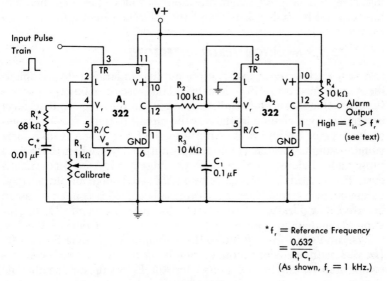

*f_r = Reference Frequency
$$= \frac{0.632}{R_t C_t}$$
(As shown, $f_r = 1$ kHz.)

Fig. 6-57. Speed alarm.

fixed 2 volts, the comparator will change states when the filtered voltage at pin 5 goes above or below 2 volts. The voltage comparison threshold becomes the voltage that corresponds to the desired frequency.

The circuit is set up for a given frequency of detection by setting a specific pulse width for monostable A_1. Since the average voltage being compared is 2 volts, which is 0.632 times the maximum voltage of 3.15 volts, a very simple criterion for the monostable pulse width evolves. Simply stated, the monostable pulse width should be 0.632 times the input pulse period. In this circuit, for example, the detection frequency is 1 kHz, which is a period of 1 ms; thus, R_t and C_t are selected to yield a 632-μs pulse width. For best results, the monostable pulse width should be trimmed for calibration, via R_1, to just actuate the comparator at the desired frequency.

This is a very sensitive comparison system, and, in practice, it can detect an "on" to "off" transition that is less than 1% of the input frequency. It can easily be set up for any frequency, up to about 100 kHz. For very low frequencies, R_3 or C_1 may have to be increased in value.

Normally, this circuit would be used to indicate an over-speed condition, or when the input frequency is greater than the reference frequency. When this occurs, the output of the circuit will go high. However, the opposite sense can also be detected by connecting the logic input of A_2 (pin 2) to the reference voltage. The output in

229

this case will be high when the input is below the reference frequency, and low when it is above the reference frequency.

6.18 POWER-MONITOR ERROR DETECTORS

A very useful circuit that is often applied in large-scale systems is the power-monitor error detector. This circuit monitors power-supply voltage(s) to determine accuracy; an incorrect voltage level is signaled by a change in state at the circuit output. Two examples of this type of circuit are shown in Fig. 6-58. Fig. 6-58A is an undervoltage monitor, which yields a high output level when the power supply goes below a predetermined voltage level. A 3905 (or 322, if desired) is used in this circuit to take advantage of its supply-independent internal reference voltage. The 3905 is connected as an inverting comparator by connecting the reference (pin 2), trigger (pin 1), and logic (pin 8) inputs together. A fraction of the V+ line is sampled by R_1 and fed to the comparator input via R_2. A 3905 (or 322 with no connection to pin 7) will have a fixed voltage-comparison threshold of 2 volts. Resistor R_1 is adjusted so that its output is just above 2 volts, with the supply just above the undervoltage limit.

In operation, for supply voltages that are above threshold, the output is low. As the supply voltage drops below threshold, the comparator changes states, with the output going high to signal an alarm condition. The high-state output can be used to drive a suitable alarm indicator.

An overvoltage monitor circuit is shown in Fig. 6-58B. This circuit is very similar to that of Fig. 6-58A, except that the 3905 is arranged as a noninverting comparator by connecting the logic input

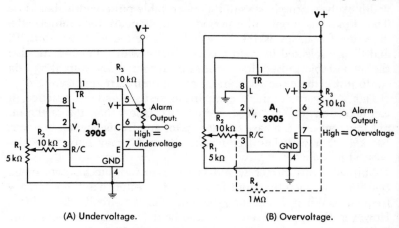

(A) Undervoltage.   (B) Overvoltage.

Fig. 6-58. Power-monitor error detectors.

(pin 8) to ground. Resistor R_1 is adjusted for an output of just below 2 volts for a supply voltage just below the desired overvoltage limit. A further increase in supply voltage will then trip the comparator, causing an alarm. In this version of the circuit some hysteresis may be desirable, and this can be achieved by using the optional resistor, R_4.

Both of these circuits are capable of good performance due to the combination of characteristics provided by the 3905 (or 322). Although both of them deliver a high-state output on an error condition, it may be desired to have a low-state (or current-sinking) output to drive a relay or alarm device, such as a Sonalert®*, directly. This can easily be accomplished by reversing the state of the logic pin from that shown in each circuit.

6.19 BURGLAR ALARM

A combination of some previously described circuits can be used as a burglar alarm sensor. This circuit is shown in Fig. 6-59. The circuit consists of a power-up monostable and a latch circuit, using both sections of a 556. When an alarm condition is sensed, the output line goes high. This signal can then be used to drive any of the alarm indicator circuits previously described.

An alarm condition is sensed when any one of the series-connected "tamper" switches is opened. These switches are strategically located at doors, windows, etc., and there can be as many of them as required. An open switch causes Q_1 to turn "on," which sets the R-S flip-flop of section B of the 556 to a high state. This constitutes an alarm condition, and once triggered, the alarm cannot be shut off by reclosing any of the tamper switches.

A device such as this needs both a delayed alarm and a reset function for operational convenience. Here, these features are provided by section A of the 556. This circuit is a 22-second power-up delay, which resets the B section of the 556 to an output-low state when power is applied or when the circuit is manually reset by S_1, the reset push-button switch. When the output of the monostable is high, the Q_1 amplifier/gate stage is disabled, which prevents triggering of the alarm. Once the delay is completed, Q_1 is enabled and the circuit is ready to be triggered when a switch is opened. A defeat switch (S_2) allows longer periods of inhibit as an additional convenience. None of the circuit components are critical, and the power supply may be any voltage within the operational parameters of the 556.

---

* A registered trademark of P. R. Mallory & Co., Inc.

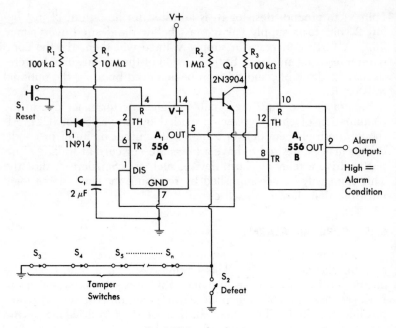

Fig. 6-59. Burglar alarm.

6.20 ALARM INDICATORS

An alarm indicator is, of course, necessary for any of the previously described alarm detectors. Two types of indicators, visual and aural, are illustrated in Figs. 6-60 and 6-61. These are activated by a high input state that matches the high-output, alarm-sensor characteristics.

A visual alarm indicator is illustrated in Fig. 6-60. This circuit is simply a gated, 555 minimum-component astable, arranged to drive an LED indicator. When the circuit is enabled, oscillations commence at a few hertz, which causes LED_1 to flash "on" and "off." The intensity is controlled by series resistor R_1, which can be adjusted for the desired brilliance with the particular supply voltage used.

As it is shown, the connection of R_1–LED_1 will cause LED_1 to be "off" on standby and to flash "on" and "off" during an alarm condition. The alternate connection to V+ will cause LED_1 to be "on" steadily at standby and to flash "on" and "off" during an alarm. For either connection, the oscillator output is also available for external use, if desired.

An aural indicator is shown in Fig. 6-61. This circuit has a combination of features designed to generate an attention-getting sound,

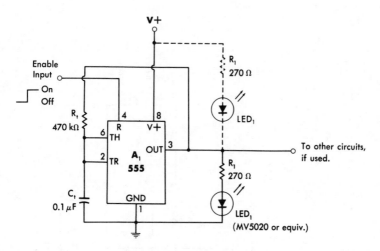

Fig. 6-60. Visual alarm indicator.

which is desirable for an alarm. There are several options available with this circuit, so it will be described in its various stages. In its simplest form, the circuit could consist of a single 555, wired as section B of the 556 as shown, with the enable input (sensor output) applied to the reset pin. This again is a minimum-component astable, with R_1 providing drive to a small PM speaker. Aural sensitivity is maximum with frequencies in the range of 1 to 2 kHz. During standby, the output will be low. When the enable line is taken high, oscillations commence, driving up to 100 mA of peak current through the speaker voice coil. This would represent a relatively high dissipation load for a 555, were it not for the fact that the duty factor is limited both by the basic astable duty cycle and the intermittent operation. Oscillation frequency is set by R_{t_2}, which is adjusted for the desired tone. For just a basic, constant tone alarm, C_1, R_2, and R_3 would not be used.

A much more attention-getting sound can be achieved by frequency-modulating or "warbling" the basic tone of this oscillator. This is achieved by using a second astable wired as section A of the 556. This is a lower-frequency oscillator whose output is superimposed on the control voltage of section B via R_2 and R_3. Capacitor C_1 smoothes the output pulses from the warble oscillator, and control R_2 determines the amount of FM, or "warble" depth, produced. Warble rate is adjusted by R_{t_1}, which controls the frequency of this oscillator. To use the circuit as a warble alarm, both oscillators should be gated by a common control line.

A third option is to add "chirp" to the sound. This is produced by gating the tone oscillator "on" only during high states of the warble

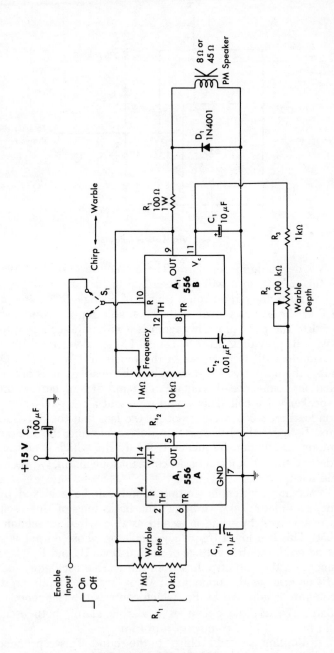

Fig. 6-61. Aural alarm indicator.

oscillator. Since the tone oscillator is also being frequency-modulated simultaneously, this produces a "chirp" sound. For this mode only, the warble oscillator is gated; gating of the tone oscillator then automatically follows.

REFERENCES

1. Alvsten, B. "Linearize Your V/F Converter." *Electronic Design,* November 8, 1973.

2. ———. "Calculate With a V/F Converter." *Electronic Design,* June 7, 1974.

3. DeFreitas, R. "Delta Sigma Modulation—The Low Cost Way to Send Data." *Electronic Design,* January 18, 1974.

4. Grebene, A. B. "Monolithic Waveform Generation." *IEEE Spectrum,* April 1972.

5. Harrison, R. "Survey of Crystal Oscillators." *Ham Radio,* March 1976.

6. Jung, W. G. "The Signal Path: Function Generators." *db, The Sound Engineering Magazine,* December 1975.

7. Kime, R. C. "The Charge Balancing A/D Converter: An Alternative to Dual Slope Integration." *Electronics,* May 24, 1973.

8. McDermott, J. "Focus on Crystals for Frequency Control." *Electronic Design,* July 5, 1976.

9. Meyer, R. G.; Sansen, M. C.; Lui, S.; Peeters, S. "The Differential Pair as a Triangle Sine Wave Converter." *IEEE Journal of Solid State Circuits,* Vol. SC-11, No. 3, June 1976.

10. Pease, R. Teledyne Inc. *Amplitude to Frequency Converter.* U.S. Patent No. 3,746,968; Filed September 1972.

11. Pouliot, F. "Have You Considered V/F Converters?" *Analog Dialogue,* Vol. 9, No. 3, 1975.

12. Robbins, M. S. *Electronic Clocks and Watches.* Howard W. Sams & Co., Inc., Indianapolis, 1975.

13. Teledyne Philbrick Applications Bulletin AN-1. *Voltage-to-Frequency Converter.* Teledyne Philbrick, Dedham, Mass.

14. Teledyne Philbrick Applications Bulletin AN-2. *Need a 1 kHz Full-Scale V-F?* Teledyne Philbrick, Dedham, Mass.

15. Teledyne Philbrick Applications Bulletin AN-6. *Magnitude-Plus-Sign ADC Using a V-F Converter.* Teledyne Philbrick, Dedham, Mass.

16. Teledyne Philbrick Applications Bulletin AN-9. *V-F's As Long-Term Integrators.* Teledyne Philbrick, Dedham, Mass.

17. Teledyne Philbrick Applications Bulletin AN-11. *V-F's F-V's and Audio Tape Recorders.* Teledyne Philbrick, Dedham, Mass.

18. Teledyne Philbrick Applications Bulletin AN-20. *Solve Your Measurement Problems With V-F and F-V.* Teledyne Philbrick, Dedham, Mass.

19. Teledyne Philbrick Applications Bulletin AN-22. *How to Specify and Test Voltage-to-Frequency and Frequency-to-Voltage Converters.* Teledyne Philbrick, Dedham, Mass.

20. Young, R. "Get ±0.02% Full-Scale VCO Accuracy." *Electronic Design,* March 15, 1973.

21. Zuch, E. L. "Voltage-to-Frequency Converters: Versatility Now at a Low Cost." *Electronics,* May 15, 1975.

III

APPENDIXES

APPENDIX **A**

Manufacturers' Data Sheets

This appendix contains catalog specification sheets for the various IC timer devices discussed in this book. The data sheets reproduced here are those of the original manufacturers.

Signetics

DESCRIPTION

The NE/SE 555 monolithic timing circuit is a highly stable controller capable of producing accurate time delays, or oscillation. Additional terminals are provided for triggering or resetting if desired. In the time delay mode of operation, the time is precisely controlled by one external resistor and capacitor. For a stable operation as an oscillator, the free running frequency and the duty cycle are both accurately controlled with two external resistors and one capacitor. The circuit may be triggered and reset on falling waveforms, and the output structure can source or sink up to 200mA or drive TTL circuits.

FEATURES

- TIMING FROM MICROSECONDS THROUGH HOURS
- OPERATES IN BOTH ASTABLE AND MONOSTABLE MODES
- ADJUSTABLE DUTY CYCLE
- HIGH CURRENT OUTPUT CAN SOURCE OR SINK 200mA
- OUTPUT CAN DRIVE TTL
- TEMPERATURE STABILITY OF 0.005% PER °C
- NORMALLY ON AND NORMALLY OFF OUTPUT

APPLICATIONS

PRECISION TIMING
PULSE GENERATION
SEQUENTIAL TIMING
TIME DELAY GENERATION
PULSE WIDTH MODULATION
PULSE POSITION MODULATION
MISSING PULSE DETECTOR

PIN CONFIGURATIONS (Top View)

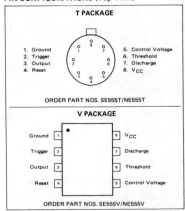

T PACKAGE

1. Ground
2. Trigger
3. Output
4. Reset
5. Control Voltage
6. Threshold
7. Discharge
8. V_{CC}

ORDER PART NOS. SE555T/NE555T

V PACKAGE

Ground	1		8	V_{CC}
Trigger	2		7	Discharge
Output	3		6	Threshold
Reset	4		5	Control Voltage

ORDER PART NOS. SE555V/NE555V

ABSOLUTE MAXIMUM RATINGS

Supply Voltage	+18V
Power Dissipation	600 mW
Operating Temperature Range	
NE555	0°C to +70°C
SE555	−55°C to +125°C
Storage Temperature Range	−65°C to +150°C
Lead Temperature (Soldering, 60 seconds)	+300°C

BLOCK DIAGRAM

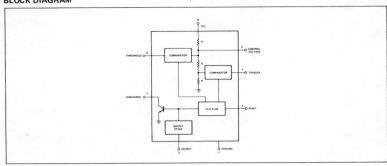

Courtesy Signetics Corp.

ELECTRICAL CHARACTERISTICS T_A = 25°C, V_{CC} = +5V to +15 unless otherwise specified

PARAMETER	TEST CONDITIONS	SE 555			NE 555			UNITS
		MIN	TYP	MAX	MIN	TYP	MAX	
Supply Voltage		4.5		18	4.5		16	V
Supply Current	V_{CC} = 5V R_L = ∞		3	5		3	6	mA
	V_{CC} = 15V R_L = ∞		10	12		10	15	mA
	Low State, Note 1							
Timing Error(Monostable)	R_A, R_B = 1KΩ to 100KΩ							
Initial Accuracy	C = 0.1 μF Note 2		0.5	2		1		%
Drift with Temperature			30	100		50		ppm/°C
Drift with Supply Voltage			0.05	0.2		0.1		%/Volt
Threshold Voltage			2/3			2/3		X V_{CC}
Trigger Voltage	V_{CC} = 15V	4.8	5	5.2		5		V
Timing Error(Astable)	V_{CC} = 5V	1.45	1.67	1.9		1.67		V
Trigger Current			0.5			0.5		μA
Reset Voltage		0.4	0.7	1.0	0.4	0.7	1.0	V
Reset Current			0.1			0.1		mA
Threshold Current	Note 3		0.1	.25		0.1	.25	μA
Control Voltage Level	V_{CC} = 15V	9.6	10	10.4	9.0	10	11	V
	V_{CC} = 5V	2.9	3.33	3.8	2.6	3.33	4	V
Output Voltage (low)	V_{CC} = 15V							
	I_{SINK} = 10mA		0.1	0.15		0.1	.25	V
	I_{SINK} = 50mA		0.4	0.5		0.4	.75	V
	I_{SINK} = 100mA		2.0	2.2		2.0	2.5	V
	I_{SINK} = 200mA		2.5			2.5		
	V_{CC} = 5V							
	I_{SINK} = 8mA		0.1	0.25				V
	I_{SINK} = 5mA					.25	.35	
Output Voltage Drop (low)	I_{SOURCE} = 200mA		12.5			12.5		
	V_{CC} = 15V							
	I_{SOURCE} = 100mA							
	V_{CC} = 15V	13.0	13.3		12.75	13.3		V
	V_{CC} = 5V	3.0	3.3		2.75	3.3		V
Rise Time of Output			100			100		nsec
Fall Time of Output			100			100		nsec

NOTES

1. Supply Current when output high typically 1mA less.

2. Tested at V_{CC} = 5V and V_{CC} = 15V.

3. This will determine the maximum value of R_A + R_B For 15V operation, the max total R = 20 megohm.

EQUIVALENT CIRCUIT (Shown for One Side Only)

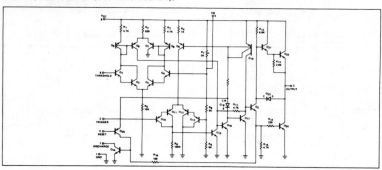

Courtesy Signetics Corp.

TYPICAL CHARACTERISTICS

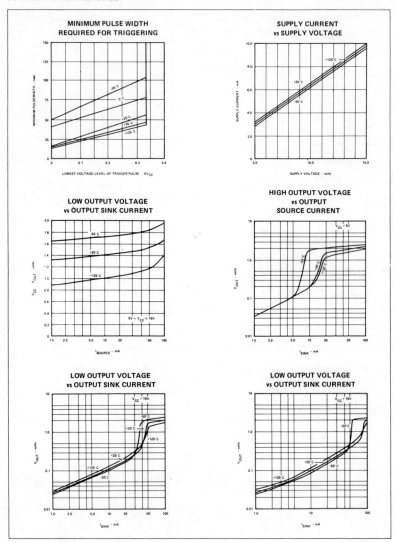

TYPICAL CHARACTERISTICS (Cont'd)

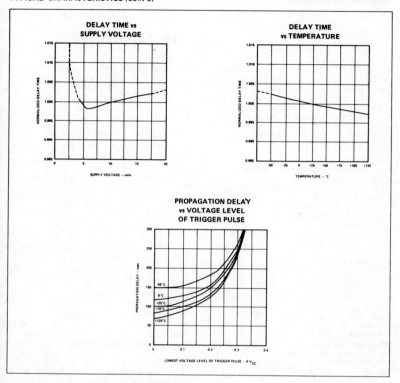

Signetics

LINEAR INTEGRATED CIRCUITS

DESCRIPTION

The NE/SE556 Dual Monolithic timing circuit is a highly stable controller capable of producing accurate time delays or oscillation. The 556 is a dual 555. Timing is provided by an external resistor and capacitor for each timing function. The two timers operate independently of each other sharing only V_{CC} and ground. The circuits may be triggered and reset on falling waveforms. The output structures may sink or source 150mA.

FEATURES

- TIMING FROM MICROSECONDS TO HOURS
- REPLACES TWO 555 TIMERS
- OPERATES IN BOTH ASTABLE, MONOSTABLE, TIME DELAY MODES
- HIGH OUTPUT CURRENT
- ADJUSTABLE DUTY CYCLE
- TTL COMPATIBLE
- TEMPERATURE STABILITY OF 0.005% PER °C

APPLICATIONS

PRECISION TIMING
SEQUENTIAL TIMING
PULSE SHAPING
PULSE GENERATOR
MISSING PULSE DETECTOR
TONE BURST GENERATOR
PULSE WIDTH MODULATION
TIME DELAY GENERATOR
FREQUENCY DIVISION
INDUSTRIAL CONTROLS
PULSE POSITION MODULATION
APPLIANCE TIMING
TRAFFIC LIGHT CONTROL
TOUCH TONE ENCODER

PIN CONFIGURATION (Top View)

A PACKAGE

Discharge	1		14	V_{CC}
Threshold	2		13	Discharge
Control Voltage	3		12	Threshold
Reset	4		11	Control Voltage
Output	5		10	Reset
Trigger	6		9	Output
Ground	7		8	Trigger

ORDER NO. SE556A, NE556A

ABSOLUTE MAXIMUM RATINGS

Supply Voltage		+18V
Power Dissipation		600mW
Operating Temperature Range	NE556	0°C to +70°C
	SE556	-55°C to +125°C
	SE556C	-55°C to +125°C
Storage Temperature Range		-65°C to +150°C
Lead Temperature (Soldering, 60 sec)		+300°C

BLOCK DIAGRAM

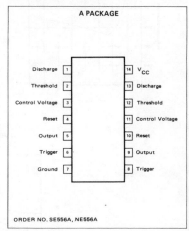

ELECTRICAL CHARACTERISTICS T_A = 25°C, V_{CC} = +5V to +15 unless otherwise specified

PARAMETER	TEST CONDITIONS	SE 556			NE 556			UNITS
		MIN	TYP	MAX	MIN	TYP	MAX	
Supply Voltage		4.5		18	4.5		16	V
Supply Current	V_{CC} = 5V R_L = ∞		3	5		3	6	mA
	V_{CC} = 15V R_L = ∞		10	11		10	14	mA
	Low State, Note 1							
Timing Error (Monostable)	R_A = 2KΩ to 100KΩ							
Initial Accuracy	C = 0.1μF Note 2		0.5	1.5		0.75		%
Drift with Temperature			30	100		50		ppm/°C
Drift with Supply Voltage			0.05	0.2		0.1		%/Volt
Timing Error (Astable)	R_A, R_B = 2KΩ to 100KΩ							
Initial Accuracy	C = 0.1μF Note 2		1.5			2.25		%
Drift with Temperature			90			150		ppm/°C
Drift with Supply Voltage			0.15			0.3		%/Volt
Threshold Voltage			2/3			2/3		X V_{CC}
Threshold Current	Note 3		30	100		30	100	nA
Trigger Voltage	V_{CC} = 15V	4.8	5	5.2		5		V
	V_{CC} = 5V	1.45	1.67	1.9		1.67		V
Trigger Current			0.5			0.5		μA
Reset Voltage		0.4	0.7	1.0	0.4	0.7	1.0	V
Reset Current			0.1			0.1		mA
Control Voltage Level	V_{CC} = 15V	9.6	10	10.4	9.0	10	11	V
	V_{CC} = 5V	2.9	3.33	3.8	2.6	3.33	4	V
Output Voltage (low)	V_{CC} = 15V							
	I_{SINK} = 10mA		0.1	0.15		0.1	.25	V
	I_{SINK} = 50mA		0.4	0.5		0.4	.75	V
	I_{SINK} = 100mA		2.0	2.25		2.0	2.75	V
	I_{SINK} = 200mA		2.5			2.5		
	V_{CC} = 5V							
	I_{SINK} = 8mA		0.1	0.25				V
	I_{SINK} = 5mA					.25	.35	
Output Voltage (high)								
	I_{SOURCE} = 200mA V_{CC} = 15V		12.5			12.5		
	I_{SOURCE} = 100mA							
	V_{CC} = 15V	13.0	13.3		12.75	13.3		V
	V_{CC} = 5V	3.0	3.3		2.75	3.3		V
Rise Time of Output			100			100		nsec
Fall Time of Output			100			100		nsec
Discharge Leakage Current			20	100		20	100	nA
Matching Characteristics (Note 4)								
Initial Timing Accuracy			0.05	0.1		0.1	0.2	%
Timing Drift with Temperature			±10			±10		ppm/°C
Drift with Supply Voltage			0.1	0.2		0.2	0.5	%/Volt

NOTES
1. Supply current when output is high is typically 1.0ma less.
2. Tested at V_{CC} = 5V and V_{CC} = 15V.
3. This will determine the maximum value of R_A + R_B for 15V operation. The maximum total R = 20 meg-ohms.
4. Matching characteristics refer to the difference between performance characteristics of each timer section.

Courtesy Signetics Corp.

EQUIVALENT CIRCUIT (Shown for One Side Only)

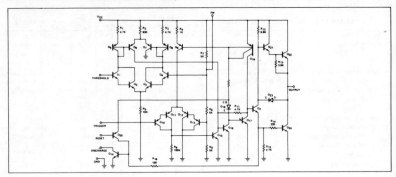

TYPICAL CHARACTERISTICS

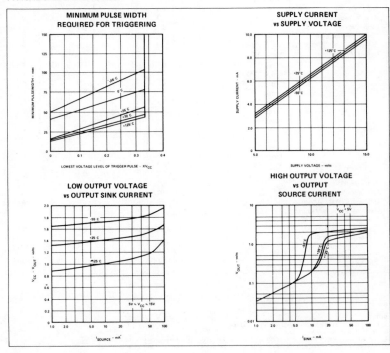

TYPICAL CHARACTERISTICS (Cont'd)

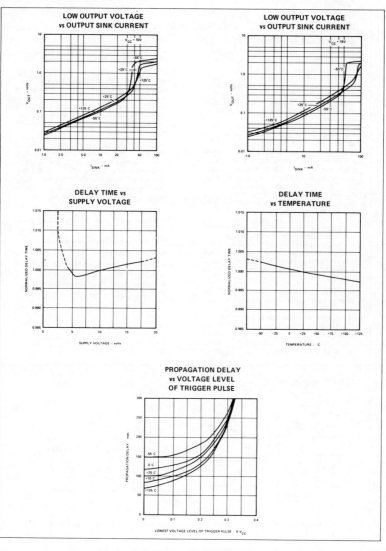

Industrial/Automotive/Functional Blocks

LM122/LM222/LM322, LM2905/LM3905 precision timers

general description

The LM122 series are precision timers that offer great versatility with high accuracy. They operate with unregulated supplies from 4.5V to 40V while maintaining constant timing periods from microseconds to hours. Internal logic and regulator circuits complement the basic timing function enabling the LM122 series to operate in many different applications with a minimum of external components.

The output of the timer is a floating transistor with built in current limiting. It can drive either ground referred or supply referred loads up to 40V and 50 mA. The floating nature of this output makes it ideal for interfacing, lamp or relay driving, and signal conditioning where an open collector or emitter is required. A "logic reverse" circuit can be programmed by the user to make the output transistor either "on" or "off" during the timing period.

The **trigger** input to the LM122 series has a threshold of 1.6V independent of supply voltage, but it is fully protected against inputs as high as ±40V — even when using a 5V supply. The circuitry reacts only to the rising edge of the trigger signal, and is immune to any trigger voltage during the timing periods.

An internal 3.15V regulator is included in the timer to reject supply voltage changes and to provide the user with a convenient reference for applications other than a basic timer. External loads up to 5 mA can be driven by the regulator. An internal 2V divider between the reference and ground sets the timing period to 1 RC. The timing period can be voltage controlled by driving this divider

with an external source through the V_{ADJ} pin. Timing ratios of 50:1 can be easily achieved.

The comparator used in the LM122 utilizes high gain PNP input transistors to achieve 300 pA typical input bias current over a common mode range of 0V to 3V. A **boost** terminal allows the user to increase comparator operating current for timing periods less than 1 ms. This lets the timer operate over a 3µs to multi-hour timing range with excellent repeatability.

The LM122 operates over a temperature range of −55°C to +125°C. An electrically identical LM222 is specified from −25°C to +85°C, and the LM322 is specified from 0°C to +70°C. The LM2905/ LM3905 are identical to the LM122 series except that the **boost** and V_{REF} pin options are not available, limiting minimum timing period to 1 ms.

features

- Immune to changes in trigger voltage during timing interval
- Timing periods from microseconds to hours
- Internal logic reversal
- Immune to power supply ripple during the timing interval
- Operates from 4.5V to 40V supplies
- Input protected to ±40V
- Floating transistor output with internal current limiting
- Internal regulated reference
- Timing period can be voltage controlled
- TTL compatible input and output

typical applications

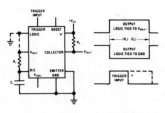

Basic Timer-Collector Output and Timing Chart

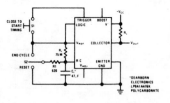

One Hour Timer with Reset and Manual Cycle End

Courtesy National Semiconductor Corp.

absolute maximum ratings

		Operating Temperature Range	
Power Dissipation	500 mW		
V⁺ Voltage	40V	LM122	$-55°C \leq T_A \leq +125°C$
Collector Output Voltage	40V	LM222	$-25°C \leq T_A \leq +85°C$
V_REF Current	5 mA	LM322	$0°C \leq T_A \leq +70°C$
Trigger Voltage	±40V	LM2905	$-40°C \leq T_A \leq +85°C$
V_ADJ Voltage (Forced)	5V	LM3905	$0°C \leq T_A \leq +70°C$
Logic Reverse Voltage	5.5V		
Output Short Circuit Duration (Note 1)			
Lead Temperature (Soldering, 10 sec)	300°C		

electrical characteristics (Note 2)

PARAMETER	CONDITIONS	LM122/LM222 MIN	TYP	MAX	LM322 MIN	TYP	MAX	LM2905/LM3905 MIN	TYP	MAX	UNITS
Timing Ratio	$T_A = 25°C, 4.5V \leq V^+ \leq 40V$	0.626	0.632	0.638	0.620	0.632	0.644	0.620	0.632	0.644	
	Boost Tied to V⁺, (Note 3)	0.620	0.632	0.644	0.620	0.632	0.644				
Comparator Input Current	$T_A = 25°C, 4.5V \leq V^+ \leq 40V$		0.3	1.0		0.3	1.5		0.5	1.5	nA
	Boost Tied to V⁺		30	100		30	100				nA
Trigger Voltage	$T_A = 25°C, 4.5V \leq V^+ \leq 40V$	1.2	1.6	2	1.2	1.6	2	1.2	1.6	2	V
Trigger Current	$T_A = 25°C, V_{TRIG} = 2V$		25			25			25		μA
Supply Current	$T_A \geq 25°C, 4.5V \leq V^+ \leq 40V$		2.5	4		2.5	4.5		2.5	4.5	mA
Timing Ratio	$4.5V \leq V^+ \leq 40V$	0.62		0.644	0.61		0.654	0.61		0.654	
	Boost Tied to V⁺	0.62		0.644	0.61		0.654				
Comparator Input Current	$4.5V \leq V^+ \leq 40V$	−5		5	−2		2	−2.5		2.5	nA
	Boost Tied to V⁺, (Note 4)			100			150				nA
Trigger Voltage	$4.5V \leq V^+ \leq 40V$	0.8		2.5	0.8		2.5	0.8		2.5	V
Trigger Current	$V_{TRIG} = 2.5V$			200			200			200	μA
Output Leakage Current	$V_{CE} = 40V$			1			5			5	μA
Capacitor Saturation Voltage	$R_t \geq 1 M\Omega$		2.5			2.5			2.5		mV
	$R_t = 10 k\Omega$		25			25			25		mV
Reset Resistance			150			150			150		Ω
Reference Voltage	$T_A = 25°C$	3	3.15	3.3	3	3.15	3.3	3	3.15	3.3	V
Reference Regulation	$0 \leq I_{OUT} \leq 3 mA$		20	50		20	50		20	50	mV
	$4.5V \leq V^+ \leq 40V$		6	25		6	25		6	25	mV
Collector Saturation Voltage	$I_L = 8 mA$		0.25	0.4		0.25	0.4		0.25	0.4	V
	$I_L = 50 mA$		0.7	1.4		0.7	1.4		0.7	1.4	V
Emitter Saturation Voltage	$T_A = 25°C, I_L = 3 mA$		1.8	2.2		1.8	2.2		1.8	2.2	V
	$T_A = 25°C, I_L = 50 mA$		2.1	3		2.1	3		2.1	3	V
Average Temperature Coefficient of Timing Ratio			0.003			0.003			0.003		%/°C
Minimum Trigger Width	$V_{TRIG} = 3V$		0.25			0.25			0.25		μs

Note 1: Continuous output shorts are not allowed. Short circuit duration at ambient temperatures up to 40°C may be calculated from $t = 120/V_{CE}$ seconds, where V_{CE} is the collector to emitter voltage across the output transistor during the short.

Note 2: These specifications apply for $T_{AMIN} \leq T_A \leq T_{AMAX}$ unless otherwise noted.

Note 3: Output pulse width can be calculated from the following equation: $t = (R_t)(C_t)\{1 - 2(0.632 - r) - V_C/V_{REF}\}$ where r is timing ratio and V_C is capacitor saturation voltage. This reduces to $t = (R_t)(C_t)$ for all but the most critical applications.

Note 4: Sign reversal may occur at high temperatures (> 100°C) where comparator input current is predominately leakage. See typical curves.

typical performance characteristics

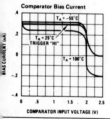

Comparator Bias Current

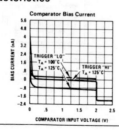

Comparator Bias Current

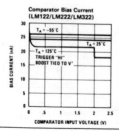

Comparator Bias Current (LM122/LM222/LM322)

typical performance characteristics (con't)

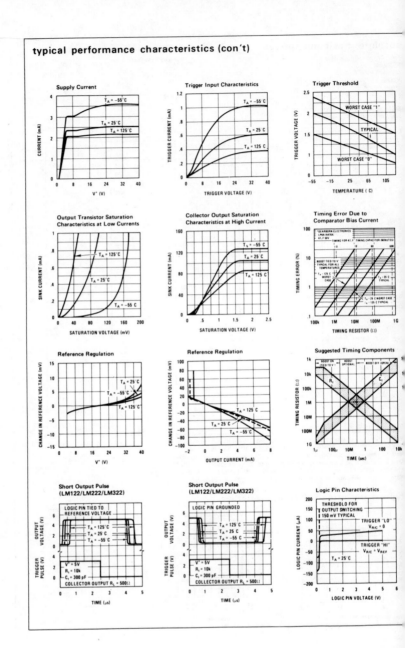

schematic diagram

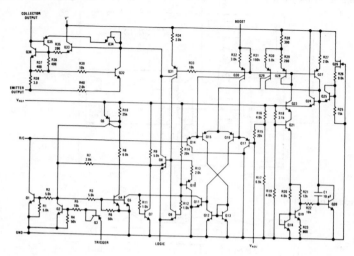

connection diagrams

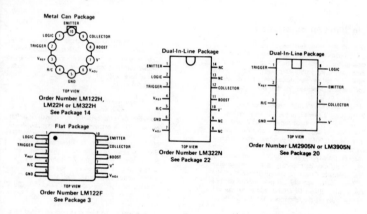

251

functional diagram

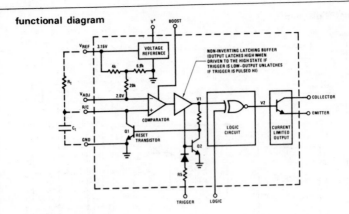

timing diagram

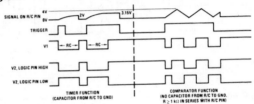

TIMER FUNCTION
(CAPACITOR FROM R/C TO GND)

COMPARATOR FUNCTION
(NO CAPACITOR FROM R/C TO GND.
R ≥ 1 kΩ IN SERIES WITH R/C PIN)

pin function description

One of the main features of the LM122 is its great versatility. Since this device is unique, a description of the functions and limitations of each pin is in order. This will make it much easier to follow the discussion of the various applications presented in this note.

V^+ is the positive supply terminal of the LM122. When using a single supply, this terminal may be driven by any voltage between 4.5V and 40V. The effect of supply variations on timing period is less than 0.005%/V, so supplies with high ripple content may be used without causing pulse width changes. Supply bypassing on V^+ is not generally needed but may be necessary when driving highly reactive loads. Quiescent current drawn from the V^+ terminal is typically 2.5 mA, independent of the supply voltage. Of course, additional current will be drawn if the reference is externally loaded.

The V_{REF} pin is the output of a 3.15V series regulator referenced to the ground pin. Up to 5.0 mA can be drawn from this pin for driving external networks. In most applications the timing resistor is tied to V_{REF}, but it need not be in situations where a more linear charging current is

required. The regulated voltage is very useful in applications where the LM122 is not used as a timer; such as switching regulators, variable reference comparators, and temperature controllers. Typical temperature drift of the reference is less than 0.01%/°C.

The **trigger** terminal is used to start a timing cycle (see functional diagram). Initially, Q1 is saturated, C_t is discharged and the latching buffer output (V1) is latched high. A trigger pulse unlatches the buffer, V1 goes low and turns Q1 off. The timing capacitor C_t connected from R/C to GND will begin to charge. When the voltage at the R/C terminal reaches the 2.0V threshold of the comparator, the buffer output (V1) in the high state. This turns on Q1, discharges the capacitor C_t and the cycle is ready to begin again.

If the **trigger** is held high as the timing period ends, the comparator will toggle and V1 will go high exactly as before. However, V1 will not be latched and the capacitor will not discharge until the trigger again goes low. When the trigger goes low, V1 remains high but is now latched.

pin function description (con't)

Trigger threshold is typically 1.6V at 25°C and has a temperature dependence of −5.0 mV/°C. Current drawn from the **trigger** source is typically 20μA at threshold, rising to 600μA at 30V, then leveling off due to FET action of the series resistor, R5. For negative input trigger voltages, the only current drawn is leakage in the nA region. The **trigger** can be driven from supplies as high as ±40V, even when device supply voltage is only 5V.

The **R/C** pin is tied to the non-inverting side of the comparator and to the collector of Q1. Timing ends when the voltage on this pin reaches 2.0V (1 RC time constant referenced to the 3.15V regulator). Q1 turns on only if the trigger voltage has dropped below threshold. In comparator or regulator applications of the timer, the trigger is held permanently high and the R/C pin acts just like the input to an ordinary comparator. The maximum voltages which can be applied to this pin are +5.5V and −0.7V. Current from the R/C pin is typically 300 pA when the voltage is negative with respect to the V_{ADJ} terminal. For higher voltages, the current drops to leakage levels. In the boosted mode, input current is typically 30 nA. Gain of the comparator is very high, 200,000 or more, depending on the state of the logic reverse pin and the connection of the output transistor.

The **ground** pin of the LM122 need not necessarily be tied to system ground. It can be connected to any positive or negative voltage as long as the supply is negative with respect to the V^+ terminal. Level shifting may be necessary for the input **trigger** if the **trigger** voltage is referred to system ground. This can be done by capacitive coupling or by actual resistive or active level shifting. One point must be kept in mind; the emitter output must not be held above the **ground** terminal with a low source impedance. This could occur, for instance, if the emitter were grounded when the **ground** pin of the LM122 was tied to a negative supply.

The terminal labeled V_{ADJ} is tied to one side of the comparator and to a voltage divider between V_{REF} and **ground**. The divider is set at 63.2% of V_{REF} with respect to ground—exactly one RC time constant. The impedance of the divider is increased to about 30k with a series resistor to present a minimum load on external signals tied to V_{ADJ}. This resistor is a pinched type with a typical variation in nominal value of −50%, +100% and a TC of 0.7%/°C. For this reason, external signals (typically a pot between V_{REF} and **ground**) connected to V_{ADJ} should have a source resistance as low as possible. For small changes in V_{ADJ}, up to several kΩ is all right, but for large variations, 250Ω or less should be maintained. This can be accomplished with a 1k pot, since the maximum impedance from the wiper is 250Ω. If a voltage is forced into V_{ADJ} from a hard source, voltage should be limited to −0.5, and +5.0V, or current limited to ±1.0 mA. This

includes capacitively coupled signals because even small values of capacitors contain enough energy to degrade the input stage if the capacitor is driven with a large, fast slewing signal. The V_{ADJ} pin may be used to abort the timing cycle. Grounding this pin during the timing period causes the timer to react just as if the capacitor voltage had reached its normal RC trigger point; the capacitor discharges and the output charges state. An exception to this occurs if the trigger pin is held high when the V_{ADJ} pin is grounded. In this case, the output changes state, but the capacitor does not discharge.

If the trigger drops while V_{ADJ} is being held low, discharge will occur immediately and the cycle will be over. If the trigger is still high when V_{ADJ} is released, the output may or may not change state, depending the voltage across the timing capacitor. For voltages below 2.0V across the timing capacitor, the output will change state immediately, then once more as the voltage rises past 2.0V. For voltages above 2.0V, no change will occur in the output. This pin is not available on the LM2905/LM3905.

In noisy environments or in comparator-type applications, a bypass capacitor on the V_{ADJ} terminal may be needed to eliminate spurious outputs because it is high impedance point. The size of the cap will depend on the frequency and energy content of the noise. A 0.1μF will generally suffice for spike suppression, but several μF may be used if the timer is subjected to high level 60 Hz EMI.

The **emitter** and the **collector** outputs of the timer can be treated just as if they were an ordinary transistor with 40V minimum collector-emitter breakdown voltage. Normally, the emitter is tied to the **ground** pin and the signal is taken from the **collector**, or the **collector** is tied to V^+ and the signal is taken from the **emitter**. Variations on these basic connections are possible. The **collector** can be tied to any positive voltage up to 40V when the signal is taken from the **emitter**. However, the **emitter** will not be pulled higher than the supply voltage on the V^+ pin. Connecting the **collector** to a voltage less than the V^+ voltage is allowed. The **emitter** should not be connected to a low impedance load other than that to which the ground pin is tied. The transistor has built-in current limiting with a typical knee current of 120 mA. Temporary short circuits are allowed; even with **collector-emitter** voltages up to 40V. The power x time product, however, must not exceed 15 watt-seconds for power levels above the maximum rating of the package. A short to 30V, for instance, can not be held for more than 4 seconds. These levels are based on 40°C maximum initial chip temperature. When driving inductive loads, always use a clamp diode to protect the transistor from inductive kick-back.

A **boost** pin is provided on the LM122 to increase the speed of the internal comparator. The comparator is normally operated at low current levels for lowest possible input current.

Courtesy National Semiconductor Corp.

pin function description (con't)

For timing periods less than 1 ms, where low input current is not needed, comparator operating current can be increased several orders of magnitude. Shorting the boost terminal to V⁺ increases the emitter current of the vertical PNP drivers in the differential stage from 25 nA to 5μA. This pin is not available on the LM2905/LM3905.

With the timer in the unboosted state, timing periods are accurate down to about 1 ms. In the boosted mode, loss of accuracy due to comparator speed is only about 800 ns, so timing periods of several microseconds can be used. The 800 ns error is relatively insensitive to temperature, so temperature coefficient of pulse width is still good.

The **Logic** pin is used to reverse the signal appearing at the output transistor. An open or "high" condition on the **logic** pin programs the output transistor to be "off" during the timing period and "on" all other times. Grounding the **logic** pin reverses the sequence to make the transistor "on" during the timing period. Threshold for the **logic** pin is typically 150 mV with 150μA flowing out of the terminal. If an active drive to the **logic** pin is desired, a saturated transistor drive is recommended, either with a discrete transistor or the open collector output of integrated logic. A maximum V_{SAT} of 75 mV at 200μA is required. Minimum and maximum voltages that may appear on the **logic** pin are 0 and +5.0, respectively.

typical applications (con't)

Basic Timers

Figure 1 is a basic timer using the collector output. R_t and C_t set the time interval with R_L as the load. During the timing interval the output may be

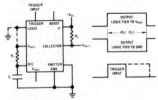

FIGURE 1. Basic Timer-Collector Output and Timing Chart

either high or low depending on the connection of the logic pin. Timing waveforms are shown in the sketch along side *Figure 1*. Note that the trigger pulse may be either shorter or longer than the output pulse width.

Figure 2 is again a basic timer, but with the output taken from the emitter of the output transistor. As with the collector output, either a high or low condition may be obtained during the timing period.

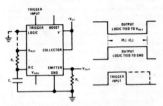

FIGURE 2. Basic Timer-Emitter Output and Timing Chart

Simulating a Thermal Delay Relay

Figure 3 is an application where the LM122 is used to simulate a thermal delay relay which

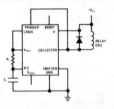

FIGURE 3. Time Out on Power Up (Relay Energized $R_t C_t$ Seconds After V_{CC} is Applied)

prevents power from being applied to other circuitry until the supply has been on for some time. The relay remains de-energized for $R_t C_t$ seconds after V_{CC} is applied, then closes and stays energized until V_{CC} is turned off. *Figure 4* is a similar circuit except that the relay is energized

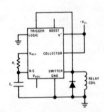

FIGURE 4. Time Out on Power Up (Relay Energized Until $R_t C_t$ Seconds After V_{CC} is Applied)

as soon as V_{CC} is applied. $R_t C_t$ seconds later, the relay is de-energized and stays off until the V_{CC} supply is recycled.

Courtesy National Semiconductor Corp.

XR-2240

Programmable Timer/Counter

PRINCIPLE OF OPERATION

The XR-2240 Programmable Timer/Counter is a monolithic controller capable of producing ultra-long time delays without sacrificing accuracy for time delays from micro-seconds up to five days. Two timing circuits can be cascaded to generate time delays up to three years. The circuit is comprised of an internal time-base oscillator, a programmable 8-bit counter and a control flip-flop. The time delay is set by an external R-C network and can be programmed to any value from 1 RC to 255 RC.

In astable operation, the circuit can generate 256 separate frequencies or pulse-patterns from a single RC setting and can be synchronized with external clock signals. Both the control inputs, pins 10-11, and the outputs, pins 1-8, are compatible with TTL and DTL logic levels.

The timing cycle for the XR-2240 is initiated by applying a positive-going trigger pulse to pin 11. The trigger input actuates the time-base oscillator, enables the counter section, and sets all the counter outputs to "low" state. The time-base oscillator generates timing pulses with its period, T, equal to 1 RC. These clock pulses are counted by the binary counter section. The timing cycle is completed when a positive-going reset pulse is applied to pin 10.

In most timing applications, one or more of the counter outputs are connected back to the reset terminal. In this manner, the circuit will start timing when a trigger is applied and will automatically reset itself to complete the timing cycle when a programmed count is completed. If none of the counter outputs are connected back to the reset terminal, the circuit would operate in its astable or free-running mode, subsequent to a trigger input.

FEATURES

Timing from micro-seconds to days
Programmable delays: 1 RC to 255 RC
Wide supply range: 4V to 15V
TTL and DTL compatible outputs
High accuracy: 0.5%
External Sync and Modulation Capability
Excellent Supply Rejection: 0.2%/V

APPLICATIONS

Precision Timing
Long Delay Generation
Sequential Timing
Binary Pattern Generation
Frequency Synthesis
Pulse Counting/Summing
A/D Conversion
Digital Sample and Hold

ABSOLUTE MAXIMUM RATINGS

Supply Voltage	18V
Power Dissipation	
Ceramic Package	750 mW
Derate above +25°C	6 mW/°C
Plastic Package	625 mW
Derate above +25°C	5.0 mW/°C
Storage Temperature	−65°C to +150°C

AVAILABLE TYPES

Part Number	Package (16 Pin DIP)	Operating Temperature
XR-2240M	Ceramic	−55°C to +125°C
XR-2240N	Ceramic	0°C to +75°C
XR-2240P	Plastic	0°C to +75°C
XR-2240CN	Ceramic	0°C to +75°C
XR-2240CP	Plastic	0°C to +75°C

SIMPLIFIED SCHEMATIC DIAGRAM

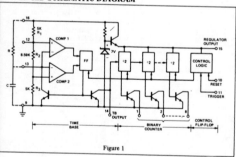

Figure 1

FUNCTIONAL BLOCK DIAGRAM

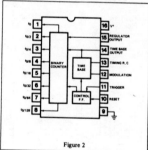

Figure 2

Courtesy Exar Integrated Systems, Inc.

ELECTRICAL CHARACTERISTICS

Test Conditions: See Figure 3, $V^+ = 5V$, $T_A = 25°C$, $R = 10\ k\Omega$, $C = 0.1\ \mu F$, unless otherwise noted.

PARAMETERS	XR-2240			XR-2240C			UNIT	CONDITIONS
	MIN	TYP	MAX	MIN	TYP	MAX		
GENERAL CHARACTERISTICS								
Supply Voltage	4		15	4		15	V	For $V^+ < 4.5V$, Short Pin 15 to Pin 16
Supply Current		3.5	6		4	7	mA	$V^+ = 5V$, $V_{TR} = 0$, $V_{RS} = 5V$
Total Circuit		12	16		13	18	mA	$V^+ = 15V$, $V_{TR} = 0$, $V_{RS} = 5V$
Counter Only		1			1.5		mA	See Figure 4
Regulator Output, V_R	4.1	4.4		3.9	4.4		V	Measured at Pin 15, $V^+ = 5V$
	6.0	6.3	6.6	5.8	6.3	6.8	V	$V^+ = 15V$, See Figure 5
TIME BASE SECTION								See Figure 3
Timing Accuracy *		0.5	2.0		0.5	5	%	$V_{RS} = 0$, $V_{TR} = 5V$
Temperature Drift		150	300		200		ppm/°C	$V^+ = 5V$ $0°C \le T \le 75°C$
		80			80		ppm/°C	$V^+ = 15V$
Supply Drift		0.05	0.2		0.08	.0.3	%/V	$V^+ \ge 8$ Volts
Max Frequency	100	130			130		kHz	$R = 1\ k\Omega$, $C = 0.007\ \mu F$
Modulation Voltage Level	3.00	3.50	4.0	2.80	3.50	4.20	V	$V^+ = 5V$
		10.5			10.5		V	$V^+ = 15V$
Recommended Range of Timing Components								See Figure 6
Timing Resistor, R	0.001		10	0.001		10	MΩ	
Timing Capacitor, C	0.007		1000	0 01		1000	μF	
TRIGGER/RESET CONTROLS								Measures at Pin 11, $V_{RS} = 0$
Trigger								
Trigger Threshold		1.4	2.0		1.4	2.0	V	
Trigger Current		8			10		μA	$V_{RS} = 0$, $V_{TR} = 2V$
Impedance		25			25		kΩ	
Response Time **		1			1		μsec	
Reset								
Reset Threshold		1.4	2.0		1.4	2.0	V	
Reset Current		8			10		μA	$V_{TR} = 0$, $V_{RS} = 2V$
Impedance		25			25		kΩ	
Response Time **		0.8			0.8		μsec	
COUNTER SECTION								See Figure 5, $V^+ = 5V$
Max. Toggle Rate	0.8	1.5			1.5		MHz	$V_{RS} = 0$, $V_{TR} = 5V$ Measured at Pin 14
Input:								
Impedance		20			20		kΩ	
Threshold	1.0	1.4		1.0	1.4		V	Measured at Pins 1 thru 8
Output:								$R_L = 3K\Omega$, $C_L = 10$ pF
Rise Time		180			180		nsec.	
Fall Time		180			180		nsec.	
Sink Current	3	5		2	4		mA	$V_{OL} \le 0.4V$
Leakage Current		0.01	8		0.01	15	μA	$V_{OH} \le 15V$

*Timing error solely introduced by XR-2240, measured as % of ideal time-base period of $T = 1.00\ RC$.
**Propagation delay from application of trigger (or reset) input to corresponding state change in counter output at pin 1.

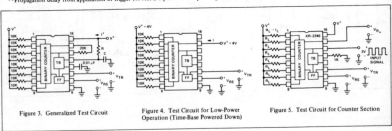

Figure 3. Generalized Test Circuit

Figure 4. Test Circuit for Low-Power Operation (Time-Base Powered Down)

Figure 5. Test Circuit for Counter Section

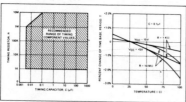

Figure 6. Recommended Range of   Figure 7. Temperature Drift of
Timing Component Values   Time-Base Period, T

DESCRIPTION OF CIRCUIT CONTROLS

COUNTER OUTPUTS (PINS 1 THROUGH 8)

The binary counter outputs are buffered "open-collector" type stages, as shown in Figure 1. Each output is capable of sinking $\approx$ 5 mA of load current. At reset condition, all the counter outputs are at high or non-conducting state. Subsequent to a trigger input, the outputs change state in accordance with the timing diagram of Figure 8.

The counter outputs can be used individually, or can be connected together in a "wired-or" configuration.

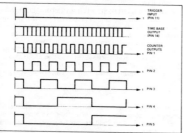

Figure 8. Timing Diagram of Output Waveforms

The combined output will be "low" as long as any one of the outputs is low. In this manner, the time delays associated with each counter output can be *summed* by simply shorting them together to a common output bus as shown in Figure 9. For example, if only pin 6 is connected to the output and the rest left open, the total duration of the timing cycle, T_O, would be 32T. Similarly, if pins 1, 5, and 6 were shorted to the output bus, the total time delay would be $T_O = (1+16+32) T = 49T$.

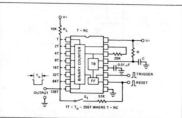

Figure 9. Circuit Connection for Timing Applications (Switch S_1 Open for Astable Operations, Closed for Monostable Operations)

RESET AND TRIGGER INPUTS (PINS 10 AND 11)

The circuit is reset or triggered with positive-going control pulses applied to pins 10 and 11. The threshold level for these controls is approximately two diode drops ($\approx$ 1.4V) above ground.

Minimum pulse widths for reset and trigger inputs, minimum trigger delay time and minimum re-trigger delay time are shown in Figure 10. Once triggered, the circuit is immune to additional trigger inputs until the end of the timing cycle.

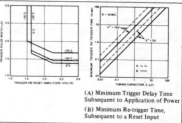

(A) Minimum Trigger Delay Time
    Subsequent to Application of Power
(B) Minimum Re-trigger Time,
    Subsequent to a Reset Input

Figure 10A. Minimum Trigger   Figure 10B. Trigger and Retrigger
and Reset Pulse Widths at   Delay Time
Pins 10 and 11

When power is applied with no trigger or reset inputs, the circuit reverts to "reset" state. Once triggered, the circuit is immune to additional trigger inputs, until the timing cycle is completed or a reset input is applied. If both the reset and the trigger controls are activated simultaneously, trigger overrides reset.

TIMING TERMINAL (PIN 13)

The time-base period T is determined by the external R-C network connected to this pin. When the time-base is triggered, the waveform at pin 13 is an exponential ramp with a period T = 1.0 RC.

Time-base can be synchronized with *integer multiples or harmonics* of input sync frequency, by setting the time-base period, T, to be an integer multiple of the sync pulse period, T_S. This can be done by choosing the timing components R and C at pin 13 such that:

$$T = RC = (T_S/m) \text{ where}$$

m is an integer, $1 \leq m \leq 10$.

MODULATION AND SYNC INPUT (PIN 12)

The period T of the time-base oscillator can be modulated by applying a dc voltage. The time-base oscillator can be synchronized to an external clock by applying a sync pulse.

TIME-BASE OUTPUT (PIN 14)

Time-Base output is an open-collector type stage, as shown in Figure 2 and requires a 20 KΩ pull-up resistor to Pin 15 for proper operation of the circuit. At reset state, the time-base output is at "high" state. Subsequent to triggering, it produces a negative-going pulse train with a period T = RC, as shown in the diagram of Figure 8.

Time-base output is internally connected to the binary counter section and also serves as the input for the external clock signal when the circuit is operated with an external time-base.

The counter input triggers on the negative-going edge of the timing or clock pulses applied to pin 14. The trigger threshold for the counter section is $\approx$ +1.5 volts. The counter section can be disabled by clamping the voltage level at pin 14 to ground.

Courtesy Exar Integrated Systems, Inc.

Note: Under certain operating conditions such as high supply voltages ($V^+ > 7V$) and small values of timing capacitor ($C < 0.1 \mu F$) the pulse-width of the time-base output at pin 14 may be too narrow to trigger the counter section. This can be corrected by connecting a 300 pF capacitor from pin 14 to ground.

REGULATOR OUTPUT (PIN 15)

This terminal can serve as a V^+ supply to additional XR-2240 circuits when several timer circuits are cascaded (See Figure 11) to minimize power dissipation. For circuit operation with external clock, pin 15 can be used as the V^+ terminal to power-down the internal time-base and reduce power dissipation. The output current shall not exceed 10 mA.

When the internal time-base is used with $V^+ \le 4.5V$, pin 15 should be shorted to pin 16.

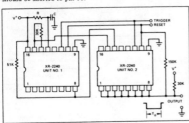

Figure 11. Low-Power Operation of Cascaded Timers

PROGRAMMING: Total timing cycle of two cascaded units can be programmed from $T_O = 256RC$ to $T_O = 65,536RC$ in 256 discrete steps by selectively shorting any one or the combination of the counter outputs from Unit 2 to the output bus.

APPLICATIONS INFORMATION

PRECISION TIMING (Monostable Operation)

In precision timing applications, the XR-2240 is used in its monostable or "self-resetting" mode. The circuit connection for this application is shown in Figure 9.

ASTABLE OPERATION

The XR-2240 can be operated in its astable or free-running mode by disconnecting the reset terminal (pin 10) from the counter outputs. Two typical circuit connections for this mode of operation are shown in Figure 12. In the circuit connection of Figure 12(a), the circuit operates in its free-running mode, with external trigger and reset signals. It will start counting and timing subsequent to a trigger input until an external reset pulse is applied. Upon application of a positive-going reset signal to pin 10, the circuit reverts back to its rest state. The circuit of Figure 12(a) is essentially the same as that of Figure 6, with the feedback switch S_1 open.

The circuit of Figure 12(b) is designed for continuous operation. The circuit self-triggering automatically when the power supply is turned on, and continues to operate in its free-running mode indefinitely.

In astable or free-running operation, each of the counter outputs can be used individually as synchronized oscillators; or they can be interconnected to generate complex pulse patterns.

OPERATION WITH EXTERNAL CLOCK

The XR-2240 can be operated with an external clock or time-base, by disabling the internal time-base oscillator and applying the external clock input to pin 14. The internal time-base can

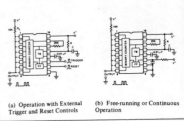

(a) Operation with External Trigger and Reset Controls (b) Free-running or Continuous Operation

Figure 12. Circuit Connections for Astable Operation

be de-activated by connecting a 1 KΩ resistor from pin 13 to ground. The counters are triggered on the negative-going edges of the external clock pulse. For proper operation, a minimum clock pulse amplitude of 3 volts is required. Minimum external clock pulse width must be $\ge 1 \mu S$.

For operation with supply voltages of 6V or less, the internal time-base section can be powered down by open-circuiting pin 16 and connecting pin 15 to V^+. In this configuration, the internal time-base does not draw any current, and the over-all current drain is reduced by ≈ 3 mA.

FREQUENCY SYNTHESIS WITH HARMONIC LOCKING: The harmonic synchronization property of the XR-2240 time-base can be used to generate a wide number of discrete frequencies from a given input reference frequency. The circuit connection for this application is shown in Figure 13. If the time base is synchronized to $(m)^{th}$ harmonic of input frequency where $1 \le m \le 10$, as described in the section on "Harmonic Synchronization", the frequency f_O of the output waveform in Figure 25 is related to the input reference frequency f_R as:

$$f_O = f_R \frac{m}{(N+1)}$$

where m is the harmonic number, and N is the programmed counter modulus. For a range of $1 \le N \le 255$, the circuit of Figure 13 can produce 1500 separate frequencies from a single fixed reference.

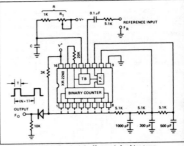

Figure 13. Frequency Synthesis by Harmonic Locking to an External Reference

One particular application of the circuit of Figure 13 is generating frequencies which are not harmonically related to a reference input. For example, by choosing the external R-C to set m = 10 and setting N = 5, one can obtain a 100 Hz output frequency synchronized to 60 Hz power line frequency.

XR-2250

BCD Programmable Timer/Counter

ADVANCE INFORMATION

The XR-2250 BCD Programmable Timer/Counter is a monolithic controller capable of producing ultra-long time delays without sacrificing accuracy. In most applications, it provides a direct replacement for mechanical or electromechanical timing devices and generates programmable time delays from micro-seconds up to 24 hours. Two timing circuits can be cascaded to generate time delays up to six months.

As shown in Fig. 1, the circuit is comprised of an internal time-base oscillator, a BCD programmable 8-bit counter and a control flip-flop. The time delay is set by an external R-C network and can be programmed to any value form 1RC to 99RC.

FEATURES

Programmable with thumb-wheel switches
Timing from micro-seconds to 24 hours
Programmable delays: 1RC to 99RC
Wide supply range: 4.5V to 15V
TTL and DTL compatible outputs
High accuracy: 0.5%
External sync and modulation capability

APPLICATIONS

Programmable timing
Long delay generation
Sequential timing
Binary pattern generation
Frequency synthesis

ABSOLUTE MAXIMUM RATINGS

Supply Voltage	18V
Power Dissipation	
Ceramic Package	750 mW
Derate above +25°C	6 mW/°C
Plastic Package	625 mW
Derate above +25°C	5.0 mW/°C
Storage Temperature	−65°C to +150°C

AVAILABLE TYPES

Part Number	Package (16 Pin DIP)	Operating Temperature
XR-2250N	Ceramic	0°C to +75°C
XR-2250P	Plastic	0°C to +75°C
XR-2250CN	Ceramic	0°C to +75°C
XR-2250CP	Plastic	0°C to +75°C

EQUIVALENT SCHEMATIC DIAGRAM

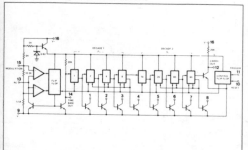

FUNCTIONAL BLOCK DIAGRAM

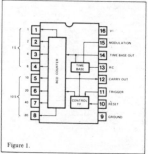

Figure 1.

Courtesy Exar Integrated Systems, Inc.

ELECTRICAL CHARACTERISTICS   Preliminary

Test Conditions: V^+ = 5V, T_A = +25°C, R = 10 KΩ, C = 0.1 μF, unless otherwise noted.

PARAMETERS	XR-2250			XR-2250C			UNITS	CONDITIONS
	MIN.	TYP.	MAX.	MIN.	TYP.	MAX.		
Supply Voltage	4.5		15	4.5		15	V	
Supply Current								
Total Circuit		5			6		mA	
Timing Accuracy*		0.5	2.0		0.5		%	V^+ = 5V
Temperture Drift		100	300		100		ppm/°C	0°C ≤ T ≤ 75°C
		80			80		ppm/°C	V^+ = 15V
Supply Drift		0.05	0.2		0.08		%/V	V^+ ≥ 8V
Max. Frequency	100	130			130		kHz	R = 1 KΩ, C = 0.007 μF
Recommended Range								
Timing Components								
Timing Resistor R	0.001		10	0.001		10	MΩ	
Timing Capacitor C	0.007		1000	0.01		1000	μF	
Trigger Threshold		1.4	2.0		1.4	2.0	V	Measured @ Pin 11
Trigger Current		8			10		μA	
Reset Threshold		1.4	2.0		1.4	2.0	V	Measured @ Pin 10
Reset Current		8			10		μA	

*Timing error solely introduced by XR-2250 measured as % of ideal time-base period of T = 1.00RC.

PRINCIPLE OF OPERATION

The timing cycle for the XR-2250 is initiated by applying a positive-going trigger pulse to Pin 11. The trigger input actuates the time-base oscillator, enables the counter section and sets all the counter outputs to "low" state. The time-base oscillator generates timing pulses with its period T, equal to 1RC. These clock pulses are counted by the BCD programmable counter section. The timing cycle is completed when a pre-set count is reached, or when a positive reset signal is applied to Pin 10.

Programming is done by selectively shorting any one or a combination of the counter outputs to the common pull-up resistor, R_L, as shown in Fig. 2. Thus, for example, to get a

time delay of 46RC, Pins 2, 3 and 7 are shorted together to give 2 + 4 + 40 = 46RC. In this manner, XR-2250 can provide 2 decades of BCD programmed timing. As shown in Fig. 3, two XR-2250 circuits can be cascaded to provide time delays from 1RC to 9999RC. In cascaded operation, carry-out terminal (pin 12) of Unit 1 is connected to counter input (pin 14) of Unit 2.

Note: To eliminate the effect of switching transients at the common output bus, it is recommended that a filter capacitor, C_L should be connected to the output, as shown in Figures 2 and 3. This filter capacitance C_L should be chosen such that the output time constant, $R_L C_L$ is ≥ 2 μsec, where R_L is the output pull-up resistor.

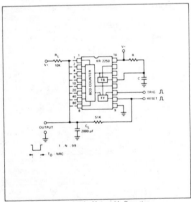

Figure 2. Circuit Connection for Monostable Operation

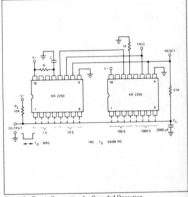

Figure 3. Circuit Connection for Cascaded Operation

Courtesy Exar Integrated Systems, Inc.

PROGRAMMABLE TIMERS/COUNTERS

Intersil

8240
8250
8260

FEATURES

- Times from microseconds to minutes, hours, or days
- Time base set by simple R, C network or external clock
- Programmable with standard thumbwheel switches
- Select output count from 1 RC to 255 RC (8240)
    - 1 RC to 99 RC (8250)
    - 1 RC to 59 RC (8260)
- Easily expanded to multiple decades (1 RC to 9,999 RC)
- Open collector outputs for flexibility
- High accuracy: ±0.5% typical
- Low drift: ±100ppm/°C typical
- Works over large supply range: 4V to 18V
- TTL compatible trigger and reset inputs

APPLICATIONS:

Programmable timing
    Process timers
    Appliance timers
    Darkroom timers
Programmable counter
    Inventory/loading/filling
    Counting/summing
Frequency generation
    Music synthesis
    Harmonic synchronization
Accurate, long-delay generator
A/D conversion
Digital Sample and Hold
Pattern generation

CONNECTION DIAGRAM

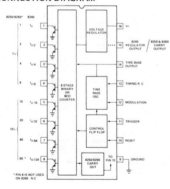

Figure 1.

GENERAL DESCRIPTION

The 8240, 8250 and 8260 are a family of monolithic programmable timer circuits. They are intended to simplify the problem of selecting various time delays or frequency outputs available from a fixed oscillator circuit.

Each device consists of an accurate, low-drift oscillator, a counter section of master-slave flip flops and appropriate logic and control circuitry all on one monolithic chip. The internal time base oscillator can be set with an external RC or can be disabled and the time base supplied from an external clock. The counter output taps are open collector transistors which can be programmed by a wire AND at external pins. Manual programming is easily accomplished by using standard thumbwheel switches. Additional logic circuitry will allow timing to be programmed by computer or microprocessor. These units are also very useful for generating ultra long delay times with relatively inexpensive RC components.

The 8260 is specifically designed to time accurate delays in seconds, minutes and hours. With its maximum count of 59 and carry out gate, a cascade of three 8260's will generate a one second clock from the 60 Hertz line, 60 seconds per minute and 60 minutes per hour programmable start to stop time. Thumbwheel switches with digits 0 to 5 and 0 to 9 are readily available to simplify the man-machine interface.

The 8250 is optimized for decimal counting and delays. It can be programmed by standard binary coded decimal (BCD) thumbwheel switches (0 to 9). Each unit gives 2 decades of counting allowing selection of time delays of from 1 RC to 99 RC. The carryout gate on the 8250 allows expansion to 9,999 or more.

The 8240 uses straight binary counting. With eight flip flops dividing down the base frequency, 8 suboctaves of the fundamental are available simultaneously in the astable mode. In the monostable mode the collectors can be wired AND to give any combination of pulse width of from 1 RC to 255 RC.

Applications for these versatile devices include appliance timers, darkroom timers and process timers. They can also be used as programmable counters. The internal clock can be disabled and the unit will count external pulses for programmable summing, loading or inventory applications. The internal clock can also be synchronized with the (m)th harmonic of an external sync and with the selectable counter, can provide a large number of non-harmonic frequencies from a single reference. Finally, they can be used as logic controlled switches in ramp type D-to-A and A-to-D converters.

ABSOLUTE MAXIMUM RATINGS

Supply Voltage	18V	
Power Dissipation		
Ceramic Package	750 mW	
Derate above +25°C	6 mW/°C	
Plastic Package	625 mW	
Derate above +25°C	5.0 mW/°C	

Operating Temperature
8240M, 8250M, 8260M   −55°C to +125°C
8240C, 8250C, 8260C   0°C to +75°C
Storage Temperature   −65°C to +150°C

ELECTRICAL CHARACTERISTICS

8240

Test Conditions: See Figure 2, V^+ = 5V, T_A = 25°C, R = 10kΩ, C = 0.1μF, unless otherwise noted.

PARAMETERS	8240M MIN.	8240M TYP.	8240M MAX.	8240C MIN.	8240C TYP.	8240C MAX.	UNITS	CONDITIONS
GENERAL CHARACTERISTICS								
Supply Voltage	4		18	4		18	V	For $V^+ < 4.5V$, Short Pin 15 to Pin 16
Supply Current								
Total Circuit (Reset)		3.5	6		4	7	mA	V^+ = 5V, V_{TR} = 0, V_{RS} = 5V
		12	16		13	18	mA	V^+ = 15V, V_{TR} = 0. V_{RS} = 5V
Total Circuit (Trigger)		24			24			V^+ = 15V, V_{TR} = 5V, V_{RS} = 0 All outputs ON. (Worst Case)
Counter Only		1			1.5		mA	See Figure 3, 8240 only
Regulator Output, V_R	4.1	4.4		3.9	4.4		V	Measured at Pin 15, V^+ = 5V
(8240 only)	6.0	6.3	6.6	5.8	6.3	6.8	V	V^+ = 15V, See Figure 4
TIME BASE SECTION								See Figure 2
Timing Accuracy		0.5	2.0		0.5	5	%	V_{RS} = 0, V_{TR} = 5V, Note 1.
Temperature Drift		150	300		200		ppm/°C	V^+ = 5V Over Operating Temperature
		80			80		ppm/°C	V^+ = 15V
Supply Drift		0.05	0.2		0.08	0.3	%/V	$V^+ \geqslant 8$ Volts, See Figure 11
Max. Frequency	100	130			130		kHz	R = 1 kΩ C = 0.007μF
Time Base Output								Measured at Pin 14
V_{TB} HIGH	2.4	2.8		2.4	2.8		V	I_{Source} = 80μA
V_{TB} LOW		0.2	0.4		0.2	0.4	V	I_{Sink} = 3.2mA
Modulation Voltage								Measured at Pin 12
Level	3.00	3.50	4.0	2.80	3.50	4.20	V	V^+ = 5V
		10.5			10.5		V	V^+ = 15V
Recommended Range								See Figure 8
of Timing Components								
Timing Resistor, R	0.001		10	0.001		10	MΩ	
Timing Capacitor, C	0.007		1000	0.01		1000	μF	
TRIGGER/RESET CONTROLS								
Trigger								Measured at Pin 11
Trigger Threshold		1.4	2.0		1.4	2.0	V	
Trigger Current		8			10		μA	V_{RS} = 0, V_{TR} = 2V
Impedance		25			25		kΩ	
Response Time		1			1		μsec.	Note 2
Reset								Measured at Pin 10
Reset Threshold		1.4	2.0		1.4	2.0	V	
Reset Current		8			10		μA	V_{TR} = 0, V_{RS} = 2V
Impedance		25			25		kΩ	
Response Time		0.8			0.8		μsec.	Note 2
COUNTER SECTION								See Figure 4, V^+ = 5V
Max. Toggle Rate	0.8	1.5			1.5		MHz	V_{RS} = 0, V_{TR} = 5V Max Input to Pin 14
Input:								
Impedance		15			15		kΩ	Measured at Pin 14
Threshold	1.0	1.4		1.0	1.4		V	
Output:								Measured at Pins 1 thru 8
Rise Time		180			180		nsec.	R_L = 3k, C_L = 10pf
Fall Time		180			180		nsec.	
V_{OUT} Low		0.2	0.4		0.2	0.4	V	I_{SINK} = 3.2 mA
Leakage Current		0.01	8		0.01	15	μA	V_{OH} = 15V

NOTE 1: Timing error solely introduced by 8240, measured as % of ideal time-base period of T = 1.00 RC.
NOTE 2: Propagation delay from application of trigger (or reset) input to corresponding state change in counter output at Pin 1.

ELECTRICAL CHARACTERISTICS

8250

Test Conditions: See Figure 2, V^+ = 5V, T_A = 25°C, R = 10 kΩ, C = 0.1 μF, unless otherwise noted.

PARAMETERS	8250M MIN.	8250M TYP.	8250M MAX.	8250C MIN.	8250C TYP.	8250C MAX.	UNITS	CONDITIONS
GENERAL CHARACTERISTICS								
Supply Voltage	4.5		18	4.5		18	V	
Supply Current								
Total Circuit (Reset)		3.5	6		4	7	mA	V^+ = 5V, V_{TR} = 0, V_{RS} = 5V
		12	16		13	18	mA	V^+ = 15V, V_{TR} = 0, V_{RS} = 5V
Total Circuit (Trigger)		24			24			V^+ = 15V, V_{TR} = 5V, V_{RS} = 0
								All outputs ON. (Worst Case)
TIME BASE SECTION								See Figure 2
Timing Accuracy		0.5	2.0		0.5	5	%	V_{RS} = 0, V_{TR} = 5V, Note 1
Temperature Drift		150	300		200		ppm/°C	V^+ = 5V Over Operating Temp.
		80			80		ppm/°C	V^+ = 15V
Supply Drift		0.05	0.2		0.08	0.3	%/V	$V^+ \geqslant$ 8 Volts, See Figure 11
Max. Frequency	100	130			130		kHz	R = 1 kΩ C = 0.007 μF
Time Base Output								Measured at Pin 14
V_{TB} HIGH	2.4	2.8		2.4	2.8		V	I_{SOURCE} = 80μA
V_{TB} LOW		0.2	0.4		0.2	0.4	V	I_{SINK} = 3.2mA
Modulation Voltage								Measured at Pin 12
Level	3.00	3.50	4.0	2.80	3.50	4.20	V	V^+ = 5V
		10.5			10.5		V	V^+ = 15V
Recommended Range								See Figure 8
of Timing Components								
Timing Resistor, R	0.001		10	0.001		10	MΩ	
Timing Capacitor, C	0.007		1000	0.01		1000	μF	
TRIGGER/RESET CONTROLS								
Trigger								Measured at Pin 11
Trigger Threshold		1.4	2.0		1.4	2.0	V	
Trigger Current		8			10		μA	V_{RS} = 0, V_{TR} = 2V
Impedance		25			25		kΩ	
Response Time		1			1		μsec.	Note 2
Reset								
Reset Threshold		1.4	2.0		1.4	2.0	V	Measured at Pin 10
Reset Current		8			10		μA	V_{TR} = 0, V_{RS} = 2V
Impedance		25			25		kΩ	
Response Time		0.8			0.8		μsec.	Note 2
COUNTER SECTION								See Figure 4, V^+ = 5V
Max. Toggle Rate	0.8	1.5			1.5		MHz	V_{RS} = 0, V_{TR} = 5V
								Max. Input Pin 14
Input:								
Inpedance		15			15		kΩ	Measured at Pin 14
Threshold	1.0	1.4		1.0	1.4		V	
Output:								Measured at Pins 1 thru 8
Rise Time		180			180		nsec.	R_L = 3k, C_L = 10 pF
Fall Time		180			180		nsec.	
V_{OUT} Low		0.2	0.4		0.2	0.4	V	I_{SINK} = 3.2 mA
Leakage Current		0.01	8		0.01	15	μA	V_{OH} = 15V
CARRY OUT GATE								See Figure 4, V^+ = 5V
V_{CO} Low		0.2	0.4		0.2	0.4	V	Measured on Pin 15
								I_{SINK} = 3.2 mA
V_{CO} High	2.4	3.5		2.4	3.5		V	I_{SOURCE} = 80μA

NOTE 1: Timing error solely introduced by 8250, measured as % of ideal time-base period of T = 1.00 RC.
NOTE 2: Propagation delay from application of trigger (or reset) input to corresponding state change in counter output at Pin 1.

Courtesy Intersil, Inc.

ELECTRICAL CHARACTERISTICS   8260

Test Conditions: See Figure 2, V^+ = 5V, T_A = 25°C, R = 10 kΩ, C = 0.1 μF, unless otherwise noted.

PARAMETERS	8260M			8260C			UNITS	CONDITIONS
	MIN.	TYP.	MAX.	MIN.	TYP.	MAX.		
GENERAL CHARACTERISTICS								
Supply Voltage	4.5		18	4.5		18	V	
Supply Current								
Total Circuit (Reset)		3.5	6		4	7	mA	V^+ = 5V, V_{TR} = 0, V_{RS} = 5V
		12	16		13	18	mA	V^+ = 15V, V_{TR} = 0, V_{RS} = 5V
Total Circuit (Trigger)		24			24			V^+ = 15V, V_{TR} = 5V, V_{RS} = 0
								All outputs ON. (Worst Case)
TIME BASE SECTION								See Figure 2
Timing Accuracy		0.5	2.0		0.5	5	%	V_{RS} = 0, V_{TR} = 5V, Note 1
Temperature Drift		150	300		200		ppm/°C	V^+ = 5V Over Operating Temp.
		80			80		ppm/°C	V^+ = 15V
Supply Drift		0.05	0.2		0.08	0.3	%/V	$V^+ \geqslant$ 8 Volts, See Figure 11
Max. Frequency	100	130			130		kHz	R = 1 kΩ, C = 0.007 μF
Time Base Output								Measured at Pin 14
V_{TB} HIGH	2.4	2.8		2.4	2.8		V	I_{Source} = 80μA
V_{TB} LOW		0.2	0.4		0.2	0.4	V	I_{Sink} = 3.2mA
Modulation Voltage								Measured at Pin 12
Level	3.00	3.50	4.0	2.80	3.50	4.20	V	V^+ = 5V
		10.5			10.5		V	V^+ = 15V
Recommended Range								See Figure 8
of Timing Components								
Timing Resistor, R	0.001		10	0.001		10	MΩ	
Timing Capacitor, C	0.007		1000	0.01		1000	μF	
TRIGGER/RESET CONTROLS								
Trigger								Measured at Pin 11
Trigger Threshold		1.4	2.0		1.4	2.0	V	
Trigger Current		8			10		μA	V_{RS} = 0, V_{TR} = 2V
Impedance		25			25		kΩ	
Response Time		1			1		μsec.	Note 2
Reset								
Reset Threshold		1.4	2.0		1.4	2.0	V	Measured at Pin 10
Reset Current		8			10		μA	V_{TR} = 0, V_{RS} = 2V
Impedance		25			25		kΩ	
Response Time		0.8			0.8		μsec.	Note 2
COUNTER SECTION								See Figure 4, V^+ = 5V
Max. Toggle Rate	0.8	1.5			1.5		MHz	V_{RS} = 0, V_{TR} = 5V
								Max Input Pin 14
Input:								
Impedance		20			20		kΩ	Measured at Pin 14
Threshold	1.0	1.4		1.0	1.4		V	
Output:								Measured at Pins 1 thru 7
Rise Time		180			180		nsec.	R_L = 3k, C_L = 10 pF
Fall Time		180			180		nsec.	
V_{OUT} Low		0.2	0.4		0.2	0.4	V	I_{SINK} = 3.2 mA
Leakage Current		0.01	8		0.01	15	μA	V_{OH} = 15V
CARRY OUT GATE								See Figure 4, V^+ = 5V
V_{CO} Low		0.2	0.4		0.2	0.4	V	Measured on Pin 15
								I_{SINK} = 3.2 mA
V_{HIGH}	2.4	3.5		2.4	3.5		V	I_{SOURCE} = 80 μA

NOTE 1: Timing error solely introduced by 8260, measured as % of ideal time-base period of T = 1.00 RC.
NOTE 2: Propagation delay from application of trigger (or reset) input to corresponding state change in counter output at Pin 1.

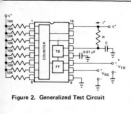

Figure 2. Generalized Test Circuit

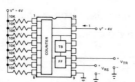

Figure 3. Test Circuit for Low-Power Operation (Time-Base Powered Down) 8240 Only.

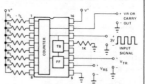

Figure 4. Test Circuit for Counter Section

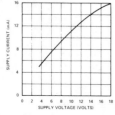

Figure 5. Supply Current vs. Supply Voltage in Reset Condition

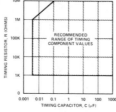

Figure 6. Recommended Range of Timing Component Values

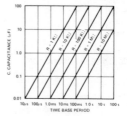

Figure 7. Time-Base Period, T, as a Function of External RC

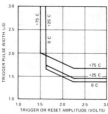

Figure 8. Minimum Trigger and Reset Pulse Widths at Pins 10 and 11

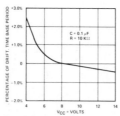

Figure 9. Power Supply Drift

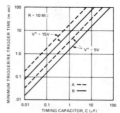

Figure 10.
A) Minimum Trigger Delay Time Subsequent to Application of Power
B) Minimum Re-trigger Time, Subsequent to a Reset Input

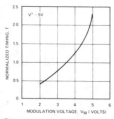

Figure 11. Normalized Change in Time-Base Period As a Function of Modulation Voltage at Pin 12

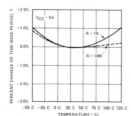

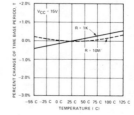

Courtesy Intersil, Inc.

ORDERING INFORMATION

TYPE	MAXIMUM COUNT	TEMPERATURE RANGE	16 PIN PACKAGE	ORDER PART NUMBER
8240C	255	0°C to +75°C	Plastic DIP	ICL 8240 C PE
8240M	255	−55°C to +125°C	Ceramic DIP	ICL 8240 M DE
8250C	99	0°C to +75°C	Plastic DIP	ICL 8250 C PE
8250M	99	−55°C to +125°C	Ceramic DIP	ICL 8250 M DE
8260C	59	0°C to +75°C	Plastic DIP	ICL 8260 C PE
8260M	59	−55°C to +125°C	Ceramic DIP	ICL 8260 M DE

PACKAGE INFORMATION

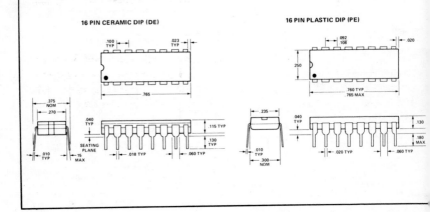

16 PIN CERAMIC DIP (DE)

16 PIN PLASTIC DIP (PE)

Courtesy Intersil, Inc.

Second-Source Guide

This appendix lists various second-source manufacturers for each IC timer device discussed in this book. The original manufacturer for each device is listed first and is printed in **bold-face** type. Addresses of the various manufacturers are listed at the end of this appendix.

Timer Type	Package	Manufacturer	Part No.
555	8-Pin Plastic MINIDIP	**Signetics**	**NE555V**
		Advanced Micro Devices	NE555V
		Exar	XR-555CP
		Fairchild	μA555TC
		Intersil	NE555V
		Lithic Systems	LS555
		Motorola	MC1455P1
		National	LM555CN
		Raytheon	RC555DN
		RCA	CA555CE
		Silicon General	SG555M
		Teledyne Semiconductor	555P
		Texas Instruments	SN72555P
556	14-Pin Plastic DIP	**Signetics**	**NE556A**
		Advanced Micro Devices	NE556A
		Exar	XR-556CP
		Fairchild	μA556PC
		Intersil	NE556A
		Lithic Systems	LS556
		Motorola	MC3556P
		National	LM556N
		Raytheon	RC556DB
		Silicon General	SG556N
		Teledyne Semiconductor	556J
322	14-Pin Plastic DIP	**National**	**LM322N**
3905	8-Pin Plastic MINIDIP	**National**	**LM3905N**
2240	16-Pin Plastic DIP	**Exar**	**XR-2240CP**
		Fairchild	μA2240PC
		Intersil	ICL8240CPE
2250	16-Pin Plastic DIP	**Exar**	**XR-2250CP***
		Intersil	ICL8250CPE*
8260	16-Pin Plastic DIP	**Intersil**	**ICL8260CPE**

* Not pin-for-pin interchangeable (see Chapter 3).

IC Timer Manufacturers

Advanced Micro Devices, Inc.
901 Thompson Place
Sunnyvale, CA 94086

Exar Integrated Systems, Inc.
750 Palomar Avenue
Sunnyvale, CA 94086

Fairchild Camera and Instrument Corp.
Semiconductor Components Group
464 Ellis Street
Mountain View, CA 94042

Intersil, Inc.
10900 North Tantau Avenue
Cupertino, CA 95014

Lithic Systems, Inc.
15800 Sanborn Road
P.O. Box 478
Saratoga, CA 95070

Motorola Semiconductor Products, Inc.
5005 East McDowell Road
Phoenix, AZ 85008

National Semiconductor Corp.
2900 Semiconductor Drive
Santa Clara, CA 95051

Raytheon Company, Semiconductor Div.
350 Ellis Street
Mountain View, CA 94042

RCA Corp., Solid State Div.
Route 202
Somerville, NJ 08876

Signetics Corp.
811 East Arques Avenue
Sunnyvale, CA 94086

Silicon General, Inc.
7382 Bolsa Avenue
Westminster, CA 92683

Teledyne Semiconductor
1300 Terra Bella Avenue
Mountain View, CA 94043

Texas Instruments, Inc.
P.O. Box 5012
Dallas, TX 75222

Timing Component Manufacturers

This appendix lists various manufacturers of timing components (capacitors and resistors) that are suitable for use with IC timers. The list is not all-inclusive, nor should it be construed as a recommendation.

Capacitors

Manufacturer	Types Manufactured
Dearborn Electronics Div. Sprague Electric Co. P.O. Box 1076 Longwood, FL 32750	Polycarbonate
Electrocube, Inc. 1710 South Del Mar Avenue San Gabriel, CA 91776	Polycarbonate Polystyrene
Midwec Div., Independent Cable P.O. Box 417G Scottsbluff, NE 69361	Polycarbonate Polystyrene
San Fernando Elec. Mfg. Co. 1501 First Street San Fernando, CA 91341	Polycarbonate Polystyrene Teflon
Siemens Corp. 186 Wood Avenue South Iselin, NJ 08830	Polycarbonate Polystyrene
Sprague Electric Co. 481 Marshall Street North Adams, MA 01247	All Types
Union Carbide Corp. Components Dept. P.O. Box 5928 Greenville, SC 29606	Paralene Tantalum

Resistors

Manufacturer	Types Manufactured
Allen-Bradley Co. Electronics Division 1201 South Second Street Milwaukee, WI 53204	General Purpose Precision Networks Trimmers & Controls
Analog Devices, Inc. P.O. Box 280 Norwood, MA 02062	Networks
Dale Electronics, Inc. P.O. Box 609 Columbus, NE 68601	Precision Semiprecision Networks Trimmers
Mepco/Electra, Inc. Columbia Road Morristown, NJ 07960	General Purpose Precision Trimmers
National Semiconductor Corp. 2900 Semiconductor Drive Santa Clara, CA 95051	Networks
Tel Labs, Inc. 154 Harvey Road Londonderry, NH 03053	Precision Networks Thermistors
Texas Instruments, Inc. P.O. Box 5012 Dallas, TX 75222	Precision Thermistors

Bibliography of IC Timer Design Ideas

This appendix lists a number of "design idea" magazine articles featuring the use of an IC timer. The articles are arranged in chronological order from 1972, the year in which the 555 was introduced, to the present. They have appeared in the following magazine columns:

Design Awards
EDN
221 Columbus Avenue
Boston, MA 02116

Designer's Casebook
Electronics
1221 Avenue of the Americas
New York, NY 10020

Ideas for Design
Electronic Design
50 Essex Street
Rochelle Park, NJ 07662

The intent of this bibliography is to serve as an aid to readers desiring additional information on the use of IC timers, and also to acknowledge the many authors who have made valuable contributions to the ever-expanding number of timer applications.

While every effort has been made toward accuracy and completeness in this bibliography, there may be some articles that were inadvertently overlooked. The author will appreciate any information of this nature for future revisions.

1972

1. Mattis, J. "Missing-Pulse Detector Reacts to 100 ns Pulse Widths." *Electronic Design*, November 23, 1972.

2. Heater, J. C. "Monolithic Timer Makes Convenient Touch Switch." *EDN*, December 1, 1972.

1973

1. James, T. W. "Single Diode Extends Duty-Cycle Range of Astable Circuit Built With Timer IC." *Electronic Design*, March 1, 1973.

2. Hoven, D. "IC Timer Makes Adjustable Schmitt Trigger." *EDN*, May 5, 1973.

3. Herring, L. W. "Timer ICs and LEDs Form Cable Tester." *Electronics*, May 10, 1973.

4. Pearl, B. "Positive Voltage Changed Into Negative, and No Transformer Is Required." *Electronic Design*, May 24, 1973.

5. Carter, J. P. "Astable Operation of IC Timers Can Be Improved." *EDN*, June 20, 1973.

6. Klement, C. "Voltage-to-Frequency Converter Constructed With Few Components Is Accurate to 0.2%." *Electronic Design*, June 21, 1973.

7. Sonsino, J. "Toroid and Photo-SCR Prevent Ground Loops in High-Isolation Biological Pulser." *Electronic Design*, June 21, 1973.

8. DeKold, D. "IC Timer Converts Temperature to Frequency." *Electronics*, June 21, 1973.

9. ———. "IC Timer Plus Thermistor Can Control Temperature." *Electronics*, June 21, 1973.

10. Harvey, M. L. "Pair of IC Timers Sounds Auto Burglar Alarm." *Electronics*, June 21, 1973.

11. McGowan, E. J. "IC Timer Automatically Monitors Battery Voltage." *Electronics*, June 21, 1973.

12. Orrel, S. A. "IC Timer Plus Resistor Can Produce Square Waves." *Electronics*, June 21, 1973.

13. Pate, J. G. "IC Timer Can Function as Low-Cost Line Receiver." *Electronics*, June 21, 1973.

14. Robbins, M. S. "IC Timer's Duty Cycle Can Stretch Over 99%." *Electronics*, June 21, 1973.

15. Black, S. L. "Get Square-Wave Tone Bursts With a Single Timer IC." *Electronic Design*, September 1, 1973.

16. Solomon, R.; Broadway, R. "DC-DC Converter Uses IC Timer." *EDN*, September 5, 1973.

17. Hefner, R. D. "Variable Speed, Synchronous Motor Control Operates From 12V Battery or 120V AC Line." *Electronic Design*, September 27, 1973.

18. Roe, B. C. "Unijunction Oscillator Helps Increase Range of Monolithic Timer Without Use of Big Capacitors." *Electronic Design*, October 25, 1973.

19. Klinger, A. R. "Integrated Timer Operates as Variable Schmitt Trigger." *Electronics*, October 25, 1973.

1974

1. Fusar, T. J. "IC Timer Makes Economical Automobile Voltage Regulator." *Electronics,* February 21, 1974.

2. Hofheimer, R. "One Extra Resistor Gives 555 Timer 50% Duty Cycle." *EDN,* March 5, 1974.

3. Miller, P. A. "Clocked Circuit Debounces Multiple Single Throw Contacts Synchronously." *Electronic Design,* March 15, 1974.

4. Blackburn, J. A. "Winking LED Notes Null for IC Timer Resistance Bridge." *Electronics,* March 21, 1974.

5. Dugan, K. R. "Making Music With IC Timers." *Electronics,* April 18, 1974.

6. Predescu, J. "Tester Built for Less Than $10 Gives Go/No-Go Check of Timer ICs." *Electronic Design,* May 24, 1974.

7. Herring, L. W. "Generating Tone Bursts With Only Two IC Timers." *Electronics,* May 30, 1974.

8. Pohlman, D. T. "Timer/Counter Chip Synthesizes Frequencies and It Needs Only a Few Extra Parts." *Electronic Design,* June 21, 1974.

9. Beckwitt, D. J. "AC Ohmmeter Provides Novel Use for Optoisolators." *EDN,* July 5, 1974.

10. Klinger, A. R. "Single Part Minimizes Differences in Monostable and Astable Periods of 555." *Electronic Design,* July 19, 1974.

11. Paiva, M. D. "Start a Logic Circuit in the Proper Mode When Power Is Turned ON or Interrupted." *Electronic Design,* August 2, 1974.

12. Laughlin, E. G. "Inexpensive Pulse Generator Is Logic Programmable." *EDN,* August 20, 1974.

13. Gartner, T. "IC Timer and Voltage Doubler Form a DC-DC Converter." *Electronics,* August 22, 1974.

14. Klinger, A. R. "Getting Extra Control Over Output Periods of IC Timer." *Electronics,* September 19, 1974.

15. Buckman, G. H. "Simple LED Flasher Is Controllable." *EDN,* September 20, 1974.

16. Gephart, R. L. "Mini-DIP Bistable Flip-Flop Sinks or Sources 200 mA." *EDN,* October 5, 1974.

17. Gulbranson, G. "Precision Timer Can Be Used to Make a Stable, Adjustable Crowbar Driver." *Electronic Design,* October 11, 1974.

18. Woodward, W. S. "Simple 10 kHz V/F Features Differential Inputs." *EDN,* October 20, 1974.

19. Martens, A. E. "Switch Selects Accurate Time Delays." *EDN,* November 5, 1974.

20. Althouse, J. "IC Timer, Stabilized by Crystal, Can Provide Subharmonic Frequencies." *Electronic Design,* November 8, 1974.

21. Morgan, L. G. "Electronic Ignition System Uses Standard Components." *Electronic Design,* November 22, 1974.

22. Felps, J. "Timer Circuit Generates Precision Power-On Reset." *Electronics,* November 28, 1974.

23. Klinger, A. R. "Generator's Duty Cycle Stays Constant Under Load." *Electronics,* November 28, 1974.

24. Long, J. D. "Burglar Alarm Is Effective Yet Simple and Inexpensive." *EDN*, December 20, 1974.

25. Galluzzi, P. "Circuit Provides Slow Auto-Wiper Cycling With 1-20 Seconds Between Sweeps." *Electronic Design*, December 20, 1974.

26. Hinkle, F. E.; Edvington, J. "Timer IC and Photocell Can Vary LED Brightness." *Electronics*, December 26, 1974.

27. Lacefield, M. M. "Simple Step-Function Generator Aids in Testing Instruments." *Electronics*, December 26, 1974.

1975

1. Dogra, S. "Operate a 555 Timer on a ±15V Supply and Deliver Op Amp Compatible Signals." *Electronic Design*, January 18, 1975.

2. Lickel, K. "Compensating the 555 Timer for Capacitance Variations." *Electronics*, February 6, 1975.

3. Kraengel, W. D. "Optically Coupled Ringer Doesn't Load Phone Line." *Electronics*, February 20, 1975.

4. Johnson, K. R. "High Voltage Power Supply From a 5V Source Regulated by Timer Feedback Circuit." *Electronic Design*, April 1, 1975.

5. Tandon, V. B. "Circuit Converts Single Trace Scope to Dual Trace Display for Logic Signals." *Electronic Design*, April 12, 1975.

6. Hinkle, F. E. "Overrange Indicator Can Enhance Frequency Meter." *Electronics*, April 17, 1975.

7. Flynn, E. A. "Put a Pendulum in Your Electronic Grandfather Clock." *EDN*, May 5, 1975.

8. Black, S. L. "555 as Switching Regulator Supplies Negative Voltage." *Electronics*, May 15, 1975.

9. Durgavich, T. "Compact DC-DC Converter Yields ±15V From +5V." *Electronics*, June 12, 1975.

10. Saunders, L. "Locked Oscillator Uses a 555 Timer." *EDN*, June 20, 1975.

11. Wise, R. M. "Capacitive Transducer Senses Tension in Muscle Fibers." *Electronics*, June 26, 1975.

12. Hilsher, R. W. "Constant Period With Variable Duty Cycle Obtained From 555 Timer With Single Control." *Electronic Design*, July 5, 1975.

13. Dance, J. B. "Ultrasonic Transmitter/Receiver Generates a 20 ft Beam That Detects Objects." *Electronic Design*, August 2, 1975.

14. Lewis, G. R. "Low-Cost Temperature Controller Built With Timer Circuits." *Electronic Design*, August 16, 1975.

15. Fleagle, J. E. "Timer ICs Control Life-Test Cycling." *Electronics*, October 2, 1975.

16. Nichols, J. C. "Versatile Delay-on-Energize Timer Uses Two 555s." *EDN*, October 5, 1975.

17. Tenny R. "Linear VCO Made From a 555 Timer." *Electronic Design*, October 11, 1975.

18. Domiciano, P. "Inverter Uses Ferrite Transformer to Eliminate Cross-Conduction." *Electronic Design*, October 25, 1975.

19. Chetty, P. R. K. "IC Timers Control DC-DC Converters." *Electronics*, November 13, 1975.

1976

1. Chetty, P. R. K. "Put a 555 Timer in Your Next Switching Regulator Design." *EDN*, January 5, 1976.

2. Gualtieri, D. M. "Triangular Waves From 555 Have Adjustable Symmetry." *Electronics*, January 8, 1976.

3. Gardner, M. R. "Line Drivers Made From 555 Timers Provide Inverted or Noninverted Outputs." *Electronic Design*, January 19, 1976.

4. Bochstabler, R. W. "Bistable Action of 555 Varies With Manufacturer." *Electronics*, February 19, 1976.

5. Jung, W. G. "Power Ramp Generator Delivers an Easily Adjustable 1A Output." *Electronic Design*, March 1, 1976.

6. Blair, D. G. "Timer Chip Becomes Meter That Detects Capacitance Changes of 1 Part in 10." *Electronic Design*, March 1, 1976.

7. Bainter, J. R. "Dual 555 Timer Circuit Restarts Microprocessor." *Electronics*, March 18, 1976.

8. Cicchiello, F. N. "Timer IC Stabilizes Sawtooth Generator." *Electronics*, March 18, 1976.

9. Hanisko, J. "Timer/Counter Functions as PLL Component." *EDN*, March 20, 1976.

10. Graf, C. R. "Audio Continuity Tester Indicates Resistance Values." *Electronics*, April 1, 1976.

11. Berlin, H. M. "555 Timer Tags Waveforms in Multiple Scope Display." *Electronics*, April 29, 1976.

12. Gergek, F. "Potentiometer and Timer Control Up/Down Counter." *Electronics*, May 13, 1976.

13. Cicchiello, F. N. "IC Timer Circuit Yields 50% Duty Cycle." *Electronics*, May 13, 1976.

14. Lo, C. C. "CD Ignition System Produces Low Engine Emissions." *EDN*, May 20, 1976.

15. Zwicker, R. M. "Phase Locked Loop Circuit Multiplies Frequencies by 2 to 256." *Electronic Design*, May 24, 1976.

16. Klinger, A. R. "Logic Probe Built From IC Timer Is Compatible With TTL, HTL & CMOS." *Electronic Design*, June 7, 1976.

17. McClellan, A. "Current Source and 555 Timer Make Linear V-F Converter." *Electronics*, June 10, 1976.

18. McNatt, M. S. "Computer Sound Effects Generated With Only Four ICs." *Electronic Design*, July 5, 1976.

19. Redmile, B. D. "Tail-Biting One-Shot Keeps Car Door Light On." *Electronics*, July 8, 1976.

20. Kranz, P.; Seger, J. "A Simple Battery Charger for Gel Cells Detects Full Charge and Switches to Float." *Electronic Design*, July 19, 1976.

21. Kraus, K. "Timer IC Paces Analog Divider." *Electronics*, August 5, 1976.

22. Jung, W. G. "555 One-Shot Circuit Features Negative Output With Positive Triggering." *Electronic Design*, August 16, 1976.

23. Sandberg, B. "State Diagrams for a 555 Timer Aid Development of New Applications." *Electronic Design*, August 16, 1976.

24. Murugesan, S. "Create a Versatile Logic Family With 555 Timers." *EDN*, September 5, 1976.

25. Lalitha, M. K.; Chetty, P. R. K. "Variable-Threshold Schmitt Trigger Uses 555 Timer." *EDN*, September 20, 1976.

26. Srinivasan, M. P. "Special-Purpose Pulse-Width Modulator Produces an Output of Same Polarity as Input." *Electronic Design*, September 27, 1976.

27. Jung, W. G. "Build a Function Generator With a 555 Timer." *EDN*, October 5, 1976.

28. Piankian, R. "Timer Extends Life of Teletypewriter." *Electronics*, November 25, 1976.

29. Sarpangal, S. "IC Timer Drives Electric Fuel Pump." *Electronics*, November 25, 1976.

30. Grundy, G. L. "Engine Staller Thwarts Car Thieves." *Electronics*, December 23, 1976.

1977

1. Morgan, D. R. "Control 10 to 10,000 Hz Digitally and Get Complementary Output Frequencies." *Electronic Design*, January 18, 1977.

2. Shiff, V. E.; Parr, R. H. "Watchdog Circuit Guards μP Systems Against Looping." *Electronic Design*, January 18, 1977.

Linear Ramp 107 monostab.

139 astab.

167 func.gen.

Index